STUDIES IN HISTORY, ECONOMICS AND PUBLIC LAW

Edited by the

FACULTY OF POLITICAL SCIENCE OF COLUMBIA UNIVERSITY

NUMBER 435

SERVICE CHARGES IN GAS AND ELECTRIC RATES

BY

HUBERT FRANK HAVLIK

SERVICE CHARGES
IN
GAS AND ELECTRIC RATES

BY

HUBERT FRANK HAVLIK, Ph.D.

INSTRUCTOR IN ECONOMICS
COLUMBIA UNIVERSITY

NEW YORK
COLUMBIA UNIVERSITY PRESS
LONDON: P. S. KING & SON, LTD.
1938

PRINTED IN THE UNITED STATES OF AMERICA

LIST OF ABBREVIATIONS

A. G. A.	American Gas Association.
E. E. I.	Edison Electric Institute.
N. A. R. U. C.	National Association of Railroad and Utilities Commissioners.
N. E. L. A.	National Electric Light Association.
P. U. R.	Public Utilities Reports.
Report of N. Y. Com. Rev. P. S. C. Law	Report of the Commission on the Revision of the Public Service Commissions Law, New York, 1930.

PREFACE

THIS study was begun at the suggestion of Professor James C. Bonbright, of Columbia University, who took an active interest in it as it progressed and whose many suggestions were of great aid to me. Several specialists have kindly discussed with me the problems raised in the course of the study. For these services I am especially indebted to Dr. John Bauer of the American Public Utilities Bureau, Mr. Frank A. Newton of the Commonwealth and Southern Corporation, and Mr. Charles S. Reed, rate consultant, now of the Duke Power Company. Many other persons occupying public or private office have graciously replied to letters of inquiry. For many suggestions made after a careful reading of the manuscript, I am beholden to Professors John Maurice Clark and Archibald H. Stockder, of Columbia University. To all these, and especially to Professor Bonbright, this work is offered in the hope that it atones for the demands made upon them.

HUBERT F. HAVLIK

TABLE OF CONTENTS

CHAPTER I

THE PROBLEM OF THE SERVICE CHARGE

A gas or electric utility company announces a new rate schedule for domestic service. Customers are told that the new rate means a " rate cut." The new schedule begins with a " service charge " of $1.00 per month for each customer; to this is added a charge for each kilowatt-hour or hundred cubic feet actually taken. The old rate did not have a service charge, and customers paid solely according to the amount of electricity or gas used at a rate per unit higher than that in the new schedule. All customers find that under the new rate they pay a service charge in addition to what they pay for energy or gas taken. It happens, therefore, that a customer who uses little service as a rule, gets larger bills than before.

This small customer asks the company to show him how his rates have been cut. He is told that only the large customers get a lower bill under the new rate than under the old. He too may get the benefit of the new rates if he consumes more than a certain amount. In reply, he insists that this is not a cut in rates for him. Why should he pay the " service charge " even when he uses little gas or electricity? All other goods and services have a standard price per unit, and when he buys them, he pays only for as much as he takes. Moreover, says he, his complaint cannot be dismissed on the ground that he is a unique case. Most of his neighbors also find their bills increased. Only the few well-to-do benefit from the new rate. To the small consumer it seems, therefore, that the company is using its monopoly power to make him, and the many of his kind, pay exorbitant rates so as to give unfair concessions to a few favored customers.

The company denies that the conclusions reached by the small customer are correct. It says that the small customer has not been paying his way, and that the service charge is the fairest and most convenient means of making him do so. Indeed, the service charge serves to remedy the discrimination against the large consumer who was overcharged under the old rates. Often, those whose bills are increased by the new schedule are well able to pay them. The new rate benefits the man of little means whose home is a center of domestic activity. He relies upon utility service to increase the amenities of home life, and to do many tasks and chores. The cut in the rate for a large consumption, which goes with the service charge, allows the use of more service at little additional cost. According to the company, the service-charge rate makes each customer pay his just share of costs, removes rate discrimination, promotes a liberal use of service by those who need it, and enables the company to collect adequate, and perhaps larger, revenues.

In the case above, both the old and new rate schedules are far more simple than those commonly in force. And the assumed " service charge " is only one of several charges that might be used instead. Nevertheless, the example does show how the service charge raises important questions. Many of these run to the root of the problem of rate making. What are the " costs " of serving a customer when he uses no service? Can these be ascertained by any sound procedure? If they can, does it follow that the small consumer ought to pay them? Are the small consumers numerous or few? Of what importance is gas and electric service to them and what are their economic characteristics? Does the service charge promote greater use of service? May not a service charge cause discrimination rather than remove it? Might it not be better to use other forms of rates?

All these questions, and many more, must be considered in making any rate schedule. It is the purpose of this study to state these questions clearly, and to attempt to answer them. Attention will be centered on residential consumers of gas and electricity, for the problem of the service charge is largely a problem of domestic rates.

The Meaning of the Term "Service Charge"

The term "service charge" has no generally accepted meaning. There are numerous variations in the form in which it appears and in its substance—*i. e.,* the items of cost which it contains. Broadly, the service charge is a fixed charge that is imposed in addition to the charge for the consumption registered upon a meter.[1] Usually it applies equally to all customers (or meters) in a given class. Sometimes, however, it varies from customer to customer in accord with variations in the costs which it is designed to cover. Ordinarily, the service charge is intended to cover customer costs, which may be defined provisionally as those traceable directly to the fact of each customer's continuing connection. Sometimes, however, it also includes a part of demand costs, which are those incurred by the company in order to be ready at all times to meet the maximum "demand" of customers.[2] Though the trend seems to be toward a service charge based on customer costs alone, many combinations of the two groups of costs may be used to make up a service charge. The great diversity of practice will be reviewed and analyzed later, in an attempt to bring some consistency into this field.

Other types of charges may be used instead of the explicit service charge. The most common of these is the minimum charge. This disappears in the consumer's bill if his use is more

1 Other uses which are not of consequence here are: a fee collected by a holding or management company for services to an operating company; an extra fare on especially fast or luxurious trains; a charge for adjusting or repairing domestic appliances; an installation or service connection charge. Outside the utility field, the service charge crops up frequently in many different uses, for examples of which see pp. 97-8, *infra.*

2 "Demand" in rate terminology means the amount of power delivered or received at a given point at any instant (known as the "load") averaged over a specified interval of time. In other words, the rate of taking service by a customer, or a group of customers. This differs from the sense in which "demand" is used in economic theory, where as a rule it refers to the quantity of goods taken in a market at a certain price. The meaning of the term when used herein is easily evident from its context. A charge based on the costs of standing ready to meet demand is called a demand charge.

than a certain minimum. For instance, a minimum charge of 50 cents, with energy at 5 cents per kw.-hr., would be absorbed in the customer's bill if he took more than 10 kw.-hr. Though the minimum charge is not like the service charge in form, it is related to the service charge in substance. If a customer uses less than the stipulated minimum, the minimum charge recovers at least a portion of customer costs.

The initial charge also is a member of the service charge family. It is a fixed minimum charge for a small first block of service. The remainder of the schedule consists of a charge for additional consumption, and infrequently, a demand charge. For example, $1.00 per month for the first 400 cu. ft. of gas or less, plus a commodity charge of 10c. per 100 cu. ft. for all additional gas, with the proviso that no bill is to be less than $1.00. The initial charge contains a hidden service charge. In the rate just cited, the per-unit cost of the gas in the initial block is so high compared with that in the following block, that the initial charge of $1.00 is not merely a minimum charge for which an equivalent amount of gas (1000 cu. ft.) would be allowed. Nor is the initial charge strictly a charge for the first block of a block-rate schedule, because the service in the initial block is negligible in comparison to that in subsequent blocks. In substance the initial charge is a service charge, though in form it is a minimum charge that includes an allowance of service equal to the first block. A charge of this type is often called a service charge.[3] But usually, and more exactly, it is called an initial charge, or initial-block charge, or initial-block-minimum charge, or minimum-block charge.[4]

3 V. M. F. Tallman, testimony, *Report of N. Y. Com. Rev. P. S. C. Law* (Albany, N. Y., 1930), III, 2275.

4 A. E. Forstall, testimony, *Report of N. Y. Com. Rev. P. S. C. Law*, III, 1918; L. H. Knapp, "House-Heating Data and Trend of Rates in New England," *Gas Age-Record*, LXIX (1932), 130; *In re* Electrical Companies (N. Y. P. S. C.), P. U. R. 1931 C 337, 348. A good term that is far too long to be convenient is "block rate with small initial charge constituting minimum monthly bill"; see Paul J. Ryan, *Some Recent Trends in Gas Rate Structures* (New York: A. G. A., 1931), *passim*.

Sometimes a rate which is based on the floor area or number of rooms in a customer's home is called a service-charge rate; for instance, an electric rate of 40c. per month for the first 500 sq. ft. of floor area or fraction thereof in a residence and 20c. for each added 500 sq. ft., plus an energy charge.[5] The area rate as expressed above is in substance a service charge. It differs from the ordinary service charge as first defined because it is not applied per meter or customer. It is best, therefore, to refer to the service charge on a room or area basis as a room-service charge or area-service charge.

The Scope of this Study

All these various types of service charges form a "family," even though some of them might also be placed in the "family" of demand charges. So far as possible this discussion will focus on rates belonging to the service charge family. A full understanding of the nature, role, and significance of service charges could be gotten only by a detailed analysis of all aspects of rate making. It is not feasible to examine here the theories and methods of allocating demand costs to various groups of customers. These costs, and the charges based on them, are treated only so far as is necessary to understand the main aspects of service charges. The same is true also of other subjects, such as commodity costs and charges, the influence of rate forms and levels on the quantity and character of consumption and revenues, and the potential market for service.

No attempt is made to face questions of fair value and fair return. Nor is any attempt made to test the operating expenses reported by utility companies to see whether or not they are necessary and reasonable. This inquiry is not concerned with the level of rates, but simply with types of schedules that are used by many companies in getting the revenues to which they are entitled by law.

5 "Report of Committee on Public Utility Rates," *N. A. R. U. C. Proceedings*, XLI (1929), 332.

The treatment of a problem often hangs on the amount and kind of facts at hand, and that is true in this case. Primary data or sources are used as a basis for discussion as much as possible. But primary data on the behavior of costs and rates in the electricity and gas supply industries are few and hard to get, and are often cast in a way that limits their usefulness. An analysis of the statistics of a single company would lead to generalizations of limited interest and validity. The main sources of data used herein are articles in technical journals, committee reports to associations, and court and commission opinions and reports. These have been combed and digested. Nevertheless, the paucity of reliable data on many aspects of this subject, and indeed on rate making in general, make it difficult to give clear answers to many of the insistent questions dealt with here.

This introduction may best conclude with a comment on the next chapter. It attempts to do two things: first, to sketch the background of the present problem of service-charge rates; second, to give an elementary account of the main types of rates for the benefit of those who are new in this field.

CHAPTER II

THE DEVELOPMENT OF SERVICE AND MINIMUM CHARGES

SERVICE and minimum charges as types of gas and electric rates can be fully understood only in the light of the development of rate theory and practice. The following pages, therefore, deal with this growth in summary fashion.[1]

In the early years of the gas and central electric station industries, the areas of supply were very small. The rates charged by supply systems were fixed largely by trial and error in order to compete with other electric and gas lighting companies, isolated electric plants, and other sources of light, heat, and power. Electric service was confined to lighting of incandescent or arc lamps, although some current was sold for the operation of direct-current motors. The gas industry fought to save its lighting business, but gradually lost it. The lost lighting business was replaced, however, by the new cooking load.

FLAT RATES

Though meters were known at the beginning of electric and gas service, most of the early companies did not use them because they were inaccurate and required outlays for installation and operation. The first types of rates were non-meter rates and were called flat rates. These were of two types: (1) a fixed sum per customer for a given period, though this amount was not the same for all customers; (2) a charge based on the customer's installation of lamps and appliances, independent of service consumed.[2]

1 The best book on public utility rate structures for the general reader is L. R. Nash's *Public Utility Rate Structures* (New York: McGraw-Hill Book Co., 1933). For convenient short treatments see: E. Jones and T. C. Bigham, *Principles of Public Utilities* (New York: The Macmillan Co., 1931), Chap. VII, VIII; W. E. Mosher and F. G. Crawford, *Public Utility Regulation* (New York: Harper & Bros., 1933), Chaps. XVII-XIX.

2 In England, and sometimes in this country, the term "flat rate" also

The first type of flat rate was not commonly used even in the early days of gas and electric service. It was unjust, for it was not related in any exact way to the capacity of the customer's appliances nor to his volume of consumption. It did not stop waste of service, for the consumer had no reason to turn out lights except to avoid the cost of replacing burned-out lamps. Because of these defects this type of flat rate never was widely used.

The flat rate based on the number and size of lamps, or on like indexes of energy- or gas-consuming devices, is correctly called a flat demand rate. The bill is based on some measure of potential demand, without regard to the energy or gas consumed. Even this rate, though it was easy to apply, simple to understand, and certain to yield a sure revenue, was soon dropped as a rate for domestic consumers.[3] The main objection to the flat demand rate was that customers wasted service, as their total bill was not based upon the amount of gas or electricity used. This rate discriminated between long- and short-hour users; it required frequent inspections in order to make sure that the customers were not using installations larger than those agreed upon; and it kept customers from increasing their installations.

Introduction of Meters and Meter Rates

In the electric lighting industry, as accurate and inexpensive meters came on the market, flat rates began to be displaced by meter rates.[4] By 1900, flat rates had been generally cast aside,

means a flat rate per unit of consumption not varying with the quantity used. In the United States this type of rate is usually called a sraight-line meter rate.

3 The flat demand rate is still used for street lighting, display lighting, industrial or commercial applications, and special domestic appliance loads, for which the capacity of the installation and the hours of use can be definitely controlled, or agreed upon in advance.

4 Though meters for gas and electricity were invented early in the history of each industry, they did not come into general use for gas service until after 1850, and for electric service until after 1895. See U. S. Bureau of

although in small towns and cities they continued in vogue for several years thereafter. The trend, however, was to meter the current and to make charges for it "upon the basis of the number of units consumed, subject to varying rates of discount or to varying methods of encouraging the customers to patronize the service. . . ."[5]

The simplest meter rate was the uniform or straight-line meter rate. In this the charge per unit of consumption was constant regardless of the quantity used; for example, five cents per kw.-hr. It was easy to apply, simple to understand, and conformed with the idea that one should pay in direct proportion to the quantity of service consumed, just as one pays for shoes.

As the market for gas and electric service expanded and diversified, it became clear that the uniform meter rate was too rigid to suit the requirements of all customers. Short-hour users often did not pay all the costs they occasioned, and the losses so incurred had to be recovered from long-hour consumers if the utility were to make a profit. Prospective large consumers usually found that they could obtain light, heat, and power more cheaply by other means—in the case of electricity, by installing their own plants. Hence, uniform rates were frequently and arbitrarily revised so as to get new business and keep the old as well. A further drawback of the straight-line rate was that as the days grew shorter every autumn, bills for service mounted in direct proportion to the increase in consumption. This caused dissatisfaction among customers.

Under the pressure of obtaining off-peak business and of utilizing plant capacity as fully as possible, the straight-line rate was replaced by new types of rates. A natural modification of the uniform meter rate was the allowance of percentage discounts which increased as the consumption or the bill increased.

Census, *Central Electric Light and Power Stations, 1902*, pp. 22, 74; A. E. Knowlton, *Electric Power Metering* (New York: McGraw-Hill Book Co., 1934), pp. 95-104.

5 *Central Electric Light and Power Stations, 1902*, p. 22.

This came to be called the step meter rate. The customer paid one rate per unit for his *entire* consumption, but this rate decreased with certain steps in the amount of consumption. For instance, 10c. per kw.-hr. for all energy consumed if the total use was from 1 to 40 kw.-hr., 7 c. per kw.-hr. if the total consumption was between 41 and 100 kw.-hr., and so on.

The step meter rate recognized the fact that the cost of supply decreased as the volume of output increased, and it encouraged the use of gas and electricity. It was, however, regressive and discriminatory. If the schedule above were in effect, a customer who took 38 kw.-hr. would receive a bill for $3.80, whereas the consumer of 54 kw.-hr. would be billed for $3.78. If the former wasted enough energy to bring his total consumption up to a certain point, within the next step, his bill would be reduced. To prevent this, it was often provided that the bill should never be less for a larger consumption than for a smaller one. Though this proviso protected the smaller user from discrimination, it did not keep a larger customer from getting some added service at no cost. Because of these defects, the step rate is practically out of use today.

The weaknesses of the step rate are avoided in the block meter rate. This requires the customer to pay a certain price per unit for his first "block" of consumption (e. g., 100 kw.-hr.), and for additional blocks (which vary in size) he pays different rates, which usually, though not always, become progressively lower. The larger consumer obtains service at rates lower than those for the first block, and is induced to utilize his installation more intensively, or to apply the service to new uses. The block meter rate has come to be one of the most widely used rates, since it is understood and accepted by the public, and can be readily adjusted to variations in the cost or value of service.

First Steps in Cost Analysis

As costs of service were reduced, increasing numbers of consumers were connected to gas and electric lines. But these often turned out to be customers who used little service as a rule, and sometimes none at all. Under these conditions, meter rates were defective in one vital respect: if consumers used little or nothing, the revenues were small proportionally, even though the costs of serving each customer were not much less than before. Hence meter rentals and minimum charges came into use, and costs of service for various groups of customers began to be analyzed. It had been early demonstrated that the rate could recognize other factors than mere quantity of use, for the "meter rent" had come into use with the meter. But it was not until 1891 that Walton Clark clearly showed that rates, particularly meter rents, should be related to various costs of service.[6] He said that the cost of actually *manufacturing* gas per thousand cubic feet was constant regardless of volume.[7] The condition of uniform costs held true only up to the inlet of the service pipe extending into the consumer's property. After that, the items of cost (interest on the investment in, and the maintenance of the meter and service pipe; expenses of accounting, billing, and meeting customer complaints) were readily determinable and were "legitimately the subject of a separate charge against the customer." This fixed charge or "meter rent" would be the same to each customer only to the extent that it covered items of bookkeeping and collection. The remaining items—interest on the investment in meters and service pipes, and the cost of their maintenance—would vary with the size of the meter and with the length of the service pipe.

Clark's presentation was the first important discussion of the costs of service that are primarily attributable to customers rather than to consumption. The meter rent for gas service, previously fixed at some percentage (usually ten) of the cost

6 Walton Clark, "Meter Rents—A Question of Equity and Policy," *Amer. Gas Light Assoc. Proceedings*, IX (1891), 517-526.

7 Clark's conception of manufacturing cost was broad, and included demand costs which are now distinguished from output costs.

of the meter,[8] was shown to have a definite cost basis, broader than the fixed charges on the meter items alone.

THE HOPKINSON DEMAND RATE

In the following year (1892), Dr. John Hopkinson, an English engineer, gave his classic paper on the costs of electric service.[9] He divided the expenses of operating a power station into those which were independent of the extent to which it was used, and those which increased in proportion to the increase in use of the plant. By the analysis of data on the generation of electricity, he showed that most of the cost of production was traceable to fixed charges upon the investment, rather than to mere output. He favored a two-part charge based on this distinction between demand and output costs. In his own words, the bill should consist of "a fixed charge per quarter proportioned to the greatest rate of supply the consumer will ever take, and a charge by meter for the actual consumption." [10] Such a rate would reward the consumer whose load factor was good, and could be adjusted to recognize the diversity in the time of demand among customers.[11]

As now applied, the Hopkinson demand rate has numerous variations. Customer costs, such as those discussed by Clark but not specifically mentioned by Hopkinson, are included in the demand component. The energy charge may be a straight-line rate, but more often is a block rate. In practice there is wide variation in the determination of the demand charge, which

8 "Report of Committee on Rate Structures," *American Gas Association Proceedings*, VIII (1926), 95. (The American Gas Association will be referred to hereafter as the A. G. A.; for other abbreviations of lengthy names see p. 5.)

9 "On the Cost of Electricity Supply," *Transactions of the Junior Engineering Society*, III, Part I (1892), 1-14. The following reference is to the reprint in *The Development of Scientific Rates for Electricity Supply* (Detroit: Printed for Private Circulation by the Edison Illuminating Co. of Detroit, 1915), pp. 7-20.

10 Hopkinson, *ibid.*, p. 13.

11 "Load factor" may be defined as the ratio of the average load over a designated period to the peak load occurring in that period.

also may be blocked. As in all demand rates, the customer's maximum demand is most accurately determined by use of a maximum-demand meter. The comparatively large expense of such an instrument for small light or power customers, has made it necessary to resort to substitute methods of estimating the demand. These have been used with varying success.[12] Domestic customers have always objected to the use of an explicit demand charge, which they cannot understand. Hence the Hopkinson demand rate has been confined to power service.

Hopkinson's paper was soon followed by other studies on costs and methods of charge. In 1896, W. J. Greene, an American utility executive, analyzed electricity costs to determine the least charge to make in order to avoid losses on small customers.[13] Greene made an elaborate grouping of expenses into (1) those dependent upon demand; (2) those affected by the number of customers; (3) those affected by output; (4) basic expenses practically independent of size of the plant, the number of customers, or the amount of current supplied; (5) expenses affected by the size of meter required by an individual customer. From this analysis he derived a formula for determining the lowest charge for which any given installation could be profitably supplied. Each customer was to be charged a minimum bill based on his estimated active load. Energy was to be billed at a uniform rate, less a discount which increased on a sliding scale as the customer increased his total consumption above the stipulated minimum. Greene's discussion was an important contribution to cost analysis despite the fact that his rate never was widely used. Apparently he was the first to distinguish clearly between the several classes of costs which are now considered fundamental in rate making.

12 See p. 25, *infra.*

13 W. J. Greene, "A Method of Calculating the Cost of Furnishing Electric Current and a Way of Selling It," *Electrical World,* XXVII (1896), 222, 223. Reprinted in *The Development of Scientific Rates for Electricity Supply,* pp. 21-30.

The Wright Demand Rate

A few months after Greene's paper was made public, Arthur Wright described a modification of the Hopkinson rate which he used in Brighton, England.[14] Although Wright held to the general analysis laid down by Hopkinson and Greene, he placed greater emphasis upon the relationship between " standing charges " and plant capacity. From an empirical study of costs, Wright found that overhead costs were proportional to the maximum demand of each customer. This demand he took to be equal to the arithmetic mean of the readings of a special maximum-demand indicator in six winter months.[15] Because there was a diversity between the peaks of individual customer's demands, there was a saving in the equipment otherwise needed to supply total requirements. Wright further found that running costs, in contrast to overhead costs, were proportional to the total energy consumed. He then proceeded to work out a system of charging for service. He was convinced that it was impolitic to follow the Hopkinson method because an undisguised demand charge might deter customers from installing any other lamps than those to be used constantly. He suggested, therefore, that a fair charge could be made by " refusing to reduce the highest permissible price per unit consumed, until the individual consumer has paid off all his proportion of the year's standing charges, as determined by the indicator demand. . . ." [16] This is the basis of the so-called Wright demand rate. A charge for demand is made by applying the maximum rate for gas or energy to a certain number of hours' use of the customer's maximum demand. All consumption in excess of this amount is billed at a lower rate. To illustrate: for the first 30 hours' use of maximum demand

14 Arthur Wright, " Cost of Electricity Supply," *Minutes of Municipal Electric Association* (Whitehall, England), 1896, pp. 1-8. Reprinted in *The Development of Scientific Rates for Electricity Supply*, pp. 33-49.

15 It was unnecessary, therefore, to measure the customer's actual demand at the time of the maximum system demand each year.

16 *The Development of Scientific Rates for Electricity Supply*, p. 46.

per month, 10c. per kw.-hr.; for all excess consumption, 6c. per kw.-hr. The Wright demand rate may be modified by the use of further energy blocks at lower rates, further demand blocks, a minimum or service charge, and so on.

Essentially, the Wright rate is a two-part rate, for it embodies a distinct demand charge even though it is based solely on volume of consumption. It thus offers lower charges for long hours' use of demand. The success of the Wright rate in Brighton, where it had been first introduced in 1893, testified to the soundness of the underlying analysis and led to its adoption in the next few years by companies operating in several of the larger cities of the United States. Since then, in one form or another, the Wright rate has been widely used for electricity supply. In the gas industry its use has been rare.

A serious drawback to all demand rates is the difficulty of ascertaining the customer's demand. The cost of a demand meter or indicator has been too heavy for all except large customers.[17] Various methods have been used to estimate the demand of residential or other small customers. At first, total connected load was used as the equivalent of actual demand.[18] This was soon replaced by the *active* connected load of the customer, as fixed by sample tests.[19] But as new appliances and lamps were bought by customers, new measurements (or estimates) of the customer's active connected load had to be made. This system of estimating demand was expensive and troublesome to administer, and it discouraged customers from making

17 According to Nash the use of demand meters, or demand attachments to watt-hour meters, might double the investment in metering facilities for small customers. *P. U. Rate Structures*, p. 49.

18 Connected load here means the sum of the continuous ratings of the load consuming apparatus on a consumer's premises. For example, the connected load is 1 kw. (or 1000 watts) if there are 10 lamps, each rated at 100 watts.

19 Active connected load is equivalent to the actual maximum use made of connected load. To continue the example in footnote 18, if five 100 watt lamps were the most that were ever in use at the same time, the active connected load would be 500 watts.

installations which would increase the demand and hence the bill. The number of outlets, floor area, or the number of rooms in a customer's premises also have been used as indexes of demand. Each of the methods has had its adherents, but each has been found wanting in accuracy, or in ease of administration, or in effectiveness in stimulating consumption, or in fairness.

Another defect of the Wright rate is that only a part of the customer's demand costs are recovered by the company when he takes less than the full amount of the first block. If his consumption is zero, the customer pays nothing toward demand costs. Again, the first block is often so wide that the consumption of many customers never exceeds it. For them the rate is the same as a uniform meter rate. These objections to the Wright rate often have been overcome by the use of minimum or service charges, or by narrowing the first block.

The Doherty or Three-Part Rate

While the several elements of costs had been segregated and recognized in rate forms (as shown in the foregoing analysis), no single rate form had been developed to follow closely the basic cost divisions. This was done in 1900 by Henry L. Doherty, an American utilities operator.[20] To improve rate practice, Doherty urged that rates be carefully patterned on "readiness-to-serve" costs. These were the minimum amount of fixed charges and expenses borne by a company in giving service, and were governed by (a) the number of customers, (b) the number of meters, and (c) the maximum current demanded. This division of costs pointed to a four-part rate, composed of customer, meter, demand, and energy charges.[21]

20 Henry L. Doherty, "Equitable, Uniform and Competitive Rates," *N. E. L. A. Proceedings*, XIX (1900), 291-321. Reprinted in *The Development of Scientific Rates for Electricity Supply*, pp. 55-78. Some years later, Doherty said, "When I devised the system of charging which is either known by my name or as the 'readiness-to-serve' method, I knew nothing of Hopkinson's work." H. L. Doherty, "Charging for Electricity," *Electrical World*, L (1907), 244.

21 Originally Doherty had intended to add a fifth factor, that of distance,

To avoid too much complexity, Doherty merged the consumer and meter components into a single customer (or meter) charge, which, with the demand and energy charges, made a three-part rate.

Though the Doherty three-part rate is theoretically superior to other rate forms, it has not been widely used. Domestic customers do not understand it. Hence it is usually stated as a two-part rate, one part of which combines the demand and customer charges. For large consumers a two-part rate is as good as a three-part rate, because customer costs are small compared with the total bill. The three-part division of costs, however, has become standard in most analyses.

Later developments in cost analysis did not change the way in which rates were cast. The method used by Doherty for electric service was later extended by others to gas costs. They added the refinement of dividing demand costs into two parts—manufacturing and distribution.[22] Thus electric- and gas-rate theories came to be similar. For a long time, however, there was no uniformity in rate practices. Electric rates usually were more complex than rates for gas. Types of charges and the actual levels of charges have varied greatly because of numerous forces and factors. Among these have been custom, competition, individual fancy, public criticism, statutes, franchises, commission rulings, advances in the technology of electricity supply, changes in the organization of the industry, and the growth of new markets for gas and electric service. These cannot be reviewed here, except as they throw light upon the origin, development, and present use of service charges.

The Spread of Service and Minimum Charges

When meters were first used, customers often owned the meters on their premises. But when the same rate was charged

to give effect to the density of customer population. But he thought this too difficult to apply equitably, and too objectionable to customers. See, State Journal Printing Co. *v.* Madison Gas and Electric Co., 4 Wis. R. C. R. 501 (1910).

22 F. W. Frueauff, "Methods of Charging for Gas," *Amer. Gas Light Assoc. Proceedings,* XXI (1904), cxiii-cxxxix.

to owners as to non-owners, there was discrimination against the former. To prevent this, companies charged the non-owners a meter rent.[23] Usually the meter rental was fixed to cover the carrying charges (interest, depreciation, maintenance, and taxes) on the meter. Other items were often added, and in time the meter rent grew to include a broad group of expenses. Clark's meter rent of 1891 included costs such as meter-reading, billing, and collecting, as well as carrying charges on the service pipe or line and the meter.[24] W. J. Greene in 1896 had isolated meter and consumer costs, but he had chosen to place them in a minimum charge along with demand costs. The Doherty rate, of course, set customer costs apart in a customer charge. As the theoretical bases of meter and consumer items became clearer, the meter rent was replaced by new forms of charges.

The disappearance of the meter rent was sped by other forces. Gas and electric companies acquired all meters on their systems, partly to escape the disputes arising from the collection of meter rentals, but largely to make it easier to inspect, test, maintain, and replace all meters. The meter rent was frowned upon by public authorities and by the law. Courts often construed maximum-rate ordinances so as not to allow meter rents. For example, the Supreme Court of Kentucky held that a gas company could not make a small consumer pay a meter rent in addition to the maximum rate of $1.35 per thousand cu. ft. fixed in its charter.[25] A few states passed laws which forbade meter rents. Thus a New York law declared that "no gas light company in this state shall charge or collect rent on its gas

23 Frueauff, *op. cit.*, p. cxiv; A. G. A. Committee on Rate Fundamentals, *Better Forms of Rates for Gas* (New York, 1927), p. 10.

24 A good discussion of customer costs and meter rentals by an Englishman, which gained little attention in America, was that by N. N. Humphreys, "Both Sides of the Ledger," *Journal of Gas Lighting* (London), LXXI (1898), 22-23, 75-76.

25 Louisville Gas Co. *v.* Dulaney and Alexander, 100 Ky. 405 (1897); see also Capital Gas and Electric Co. *v.* Gaines, 20 Ky. L. 1464, 49 S. W. 462 (1899).

meters, either in a direct or indirect manner." [26] Again, some city ordinances did not allow the use of meter rents.[27] At best the meter rent was of doubtful legality.[28] Regulatory bodies often said that the meter rent was undesirable, and that it discriminated against those customers who did not own meters.[29] Where the meter rent had been added to a minimum charge which it duplicated, commissions often required the elimination of the meter charge.[30] For all these reasons meter rents were gradually deleted from rate schedules.

When meter rents were abandoned, many companies lost revenues which were hard to regain by increasing the commodity rate. Hence they turned to other sources of revenue, such as service or minimum charges. By and large, the service charge was bitterly opposed by the growing class of small consumers.[31] Because of this antagonism, the service charge did not spread rapidly. There was a transition from the meter rent to the minimum charge instead of a change from the meter rent to the service charge. Where the term " meter rent " was kept, it was usually a substitute for " minimum charge " or

26 N. Y. Laws 1890, c. 566. This runs back to Laws 1868, c. 253. Such laws were upheld by the courts in New York and in other states. See City of Buffalo *v.* Buffalo Gas Co., 81 App. Div. 505, 80 N. Y. Supp. 1093 (4th Dep't., 1903).

27 See, for instance, Tipton *v.* Tipton Light and Heat Co., 176 Iowa 224 (1916). An ordinance fixed a maximum rate, and gave the right to collect a minimum charge, but prohibited a meter rent. The Supreme Court of Iowa held that the city had power to forbid a meter rent so long as the maximum rate was reasonable without the rent.

28 B. Wyman, *Public Service Corporations* (New York: Baker, Voorhis & Co., 1921), p. 703.

29 See, for instance, City of Geneva *v.* Equitable Electric Light Co., 6 Wis. R. C. R. 203 (1913); Laclede *v.* Laclede Electric Light Co. (Mo. P. S. C.), P. U. R. 1925 E 373; Bosshard *v.* Hussa Bros. Light and Water Co. (Wis. R. C.), P. U. R. 1925 E 584.

30 For example, *In re* Ladysmith Lighting Co. (Wis. R. C.), P. U. R. 1915 A 1050; Meek *v.* Consumers Electric Light and Power Co. (Mo. P. S. C.), P. U. R. 1915 A 956.

31 S. Scovil, Discussion of H. L. Doherty's paper, *N. E. L. A. Proceedings,* XIX (1900), 333.

for "service charge." [32] It must be remembered, however, that the methods used to ascertain service charges, meter rents, and minimum charges were not standardized. There was little agreement among rate experts on the nature or the fairness of each of these charges.[33]

The advent of high-efficiency lamps for electricity showed that fixed or minimum charges assured some steady income. By 1910 the tungsten-filament lamp was fast displacing the older types of electric lamps (chiefly carbon filament), as well as gas lamps. The new lamps cut down the energy used by as much as two-thirds, and immediate revenues from lighting sales also dropped. As the cost of electric lighting decreased, large numbers of small consumers who had formerly used gas lighting asked for electric service. The companies, however, found that customer and demand costs per consumer were largely the same as before, though the latter dropped somewhat. In the case of small consumers, these two groups of costs made up most of the cost of service.

Managers were not agreed on how to meet the problems raised by the conjunction of declining revenues, stationary costs, and a growth in customers served. Some said that it would be best to follow the rate ideas of Hopkinson, Wright, and Doherty. It was argued that if this were done, the plant capacity made idle by high-efficiency lamps could be used to take care of the new business attracted by the reduced cost of lighting. This was regarded as a good opportunity for the industry to grow extensively and intensively.[34] Many companies recast their rates so as to give more weight to fixed costs. In this way the minimum charge, the customer charge, and to some extent

32 Frueauff, *op. cit.*, p. cxiv.

33 Contrast, for example, the views expressed in the following discussions: Alton D. Adams, "Meter Rents and Minimum Rates," *Electrical World*, LV (1910), 525-527; "Concerning Meter Rents" (editorial), *Electrical World*, LV (1910), 502-503.

34 S. E. Doane, "High Efficiency Lamps—Their Effect on the Cost of Light to the Central Stations," *N. E. L. A. Proceedings*, XXXVIII (1910), 152-170. Reprinted in *The Development of Scientific Rates for Electricity*

the meter rent came to be used more widely.[35] These moderate changes almost always worked out well, for both old and new customers took more electricity than before, and total revenues soon regained and even surpassed their old levels.[36] Some companies, however, thought that these mild changes did not make meter rates safe enough under the new conditions, and, therefore, returned to the old forms of flat demand rates. These usually required some limitation of maximum demand. They remained in effect but a short time, for the experience of the more progressive companies proved that meter rates, with protective fixed charges, were better than flat demand rates.[37]

In these years several states had begun to control public utilities through commissions, which were given authority over the making of rates and charges. Though this authority was limited in many ways, the commissions could, and did, require all companies under their jurisdiction to file a complete list of rates. This disclosed many discriminations and incongruities in rate practice which were hard to justify to the ratepayers. As commissions reviewed new and old rates, they were able to better the design of rate schedules. The Wisconsin Railroad Commission, for instance, made telling use of two- and three-part cost analyses in the design of demand rates.[38] As a result of this stress on cost analysis numerous utilities came to use service, minimum, or demand charges. But rate-making for gas and

Supply, pp. 81-99. Doane's analysis traced the effect of a reduction in consumption upon costs. It demonstrated why consumers' bills could not be reduced in proportion to the increase in lamp efficiency.

35 U. S. Bureau of Census, *Central Electric Light and Power Stations, 1912*, pp. 166-168.

36 "Report of Lamp Committee," *N. E. L. A. Proceedings*, XXXVIII (1910), 110-113.

37 "Report of the Committee on Residence Business," *N. E. L. A. Proceedings*, XLIII (1912), 441-442.

38 For an account of the early experience of the Wisconsin Commission with rate-making and cost analysis, see Fred L. Holmes, *Regulation of Railroads and Public Utilities in Wisconsin* (New York: D. Appleton & Co., 1915), Chap. XII.

electricity was not a scientific procedure. In the main, rates continued to reflect the forces of custom, competition, and commercial expediency.

During the World-War years, and after, the question of the form and adequacy of utility rates came to the fore. The swift rise in costs of operation led companies to ask for higher rates, and commissions often approved them. The additions to rates were horizontal, and took the form of percentage surcharges upon the bill at the old rate, or of flat additions to the output charge, or of fuel charges varying as the cost of fuel went above or below some set level.[39] Larger revenues also could be gotten by adding a service charge to the old rates, and this was often done when horizontal increases threatened to drive away customers who had other sources of supply.[40] Service charges of this kind were said to be an especially good source of revenues because they fell chiefly upon those customers who had been served at a loss.[41] For these reasons, service charges were used increasingly, especially in the gas industry.

The service charges of the post-war period frequently were a means of fitting rates to the costs of serving various groups of customers more closely than was otherwise possible.[42]

39 John W. Lieb, "The Electrical Industry in War Service," *N. E. L. A. Proceedings*, LXVIII (1918), 29; "Report of the Rate Research Committee," *N. E. L. A. Proceedings*, LXIX (1919), 23.

40 Lieb, *op. cit.*, p. 23; A. G. A. Committee on Rate Fundamentals, *op. cit.*, pp. 10-11; "Symposium on Adoption of the Service Charge," *Gas Age-Record*, XLVIII (1921), 133-136; Empire State Gas and Electric Association, *The Service Charge Rate* (New York: 1918?), p. 2; Hartford *v.* Hartford Gas Light Co. (Conn. P. U. C.), P. U. R. 1920 F 840; *In re* Gardner Gas Fuel and Light Co., 3 Mass. D. P. U. 55 (1922).

41 See, for instance, *Re* Public Service Electric and Gas Co. (N. J. B. P. U. C.), P. U. R. 1929 E 21; *In re* St. Joseph Gas Co., 10 Mo. P. S. C. R. 122 (1920). Also, "Report of the Committee on Public Utility Rates," *N. A. R. U. C. Proceedings*, XXI (1919), 68.

42 See, for instance, *Re* Utah Gas and Coke Co. (Utah P. U. C.), P. U. R. 1919 D 645; City Council *v.* Providence Gas Co. (R. I. P. U. C.), P. U. R. 1921 D 842; Municipal Gas Co. *v.* Wichita Falls (Tex. R. C.), P. U. R. 1924 B 410; Kinch et al. *v.* Concord Light and Power Co., 6 N. H. P. S. C. R. 381 (1921); "Report of Committee on Public Utility Rates," *N. A. R. U. C. Proceedings*, XXI (1919), 68.

But the use of service charges to obtain increased revenues in order to offset rising costs served to misrepresent their real nature.[43] Ratepayers came to look upon the service charge as an addition to their bills for which they received no more service.[44] Some utility managers warned their fellows that the use of a service charge for emergency purposes was dangerous as well as improper. They saw that it might bar the proper use of the service charge to correct rate defects and discriminations.[45] This warning was well founded. In later years when many companies tried to introduce permanent service charges, they usually met with bitter opposition by consumers who had come to think of service charges as rate increases.[46]

The Present-Day Importance of the Service Charge

In post-war years the problem of the small consumer continued to vex rate-makers. The number of domestic gas and electricity consumers grew continuously. The average consumption per customer often increased year by year, but the monthly

43 A vital defect of the service charge as a temporary surcharge was that it was not functionally related to customer or demand costs. A rise in the cost of fuel and labor affected output cost to a far greater extent than it did customer cost. A service charge based upon increases in fuel and labor costs bore most heavily upon domestic customers despite the fact that the cost of serving other classes of customers had increased. Some of the unpopularity of the service charge probably arose on this ground, for often protests were made against the way in which rates were increased rather than the increase as such. See, G. C. Mathews, "Round Table Discussion on 'The Gas Problem,'" *N. A. R. U. C. Proceedings*, XXII (1920), 190, 191; *Re* Coast Gas Co. (N. J. B. P. U. C.), P. U. R. 1923 D 352; Hartford *v.* Hartford Gas Light Co. (Conn. P. U. C.), P. U. R. 1920 F 840.

44 Though emergency rates were not supposed to enlarge the fair rate of return, some attempts to get exorbitant returns were screened by the use of the service charge. See, J. E. Bullard, "Selling the Service Charge," *American Gas Journal*, CXV (1921), 278; City of Meriden *v.* Meriden Gas Light Co. (Conn. P. U. C.), P. U. R. 1921 B 611.

45 Lieb, *op. cit.*, p. 23; Empire State Gas and Electric Assoc., *op. cit.*, p. 2.

46 Helen E. C. Farnsworth, *Rate Structures in the Manufactured Gas Industry* (Doctoral Dissertation filed at Stanford University, 1930), p. 31; C. S. Reed, "Installing Better Forms of Rates for Gas," *A. G. A. Monthly*, IX (1927), 561-562.

or annual consumption of large numbers of domestic customers stayed very low. Forces were at work which restrained and cut down the annual consumption of electricity, and to a greater extent, of gas.[47] The mode of life in cities was swiftly changing. This was shown by the trend toward small homes and apartments, and in the dwindling of the urban family. In many homes less and less use was made of kitchens and cooking equipment as ever-growing quantities of bakery, canned, and prepared foods were bought for home consumption, and as the habit of eating in restaurants became common. The chores of laundering and ironing were often shifted to the prospering central laundries.[48] In multi-family dwellings, hot water and heat were supplied by the landlord, thus cutting off a market for gas or electric service.

These changes limited the use of gas and electricity in many homes. Yet the customer costs incident to each consumer ran on. These costs were said to be so large that, under a simple meter rate, the small consumer did not pay enough to cover them. The resultant loss had to be shifted to larger consumers if the company were not to suffer a deficit in the fair return. It was said, therefore, that the rate used should be one which would reimburse the utility for all of the customer costs, or for as much of them as possible. The simplest rate of that kind was one with a minimum or service charge.

47 See, H. H. Agee, "Effect of Economic Conditions and Social Trends on Rates," *Gas Age-Record,* LXIII (1929), 643-646, 649, and "The Problem of the Small Consumer," *P. U. Fortnightly,* V (1930), 670-671; The Presidents' Research Committee on Social Trends, *Recent Social Trends* (New York: McGraw-Hill Book Co., 1933), pp. 473-476, 664-668, 671, 865, 897-900; *Re* Public Service Electric and Gas Co. (N. J. B. P. U. C.), P. U. R. 1929 E 17.

48 The number of gas meters installed in the territory of the Brooklyn Union Gas Company increased by 22% from 1921 to 1928, whereas, in the same period, commercial laundry establishments increased by 116%, restaurants, by 199%, and delicatessen stores, by 138%. See, *In re* Tariff Schedules for Gas Proposed Effective May 16, 1928, of the Brooklyn Union Gas Co., Case No. 4832, N. Y. P. S. C., Exhibit No. 45.

Rates of the service-charge type were found good on other grounds. During the 1920's utility managers came to see that domestic use of gas and electricity, long limited chiefly to gas cooking and electric lighting, could be enlarged if new and old types of domestic appliances were used commonly. New rates were needed to make the use of gas or electricity for mechanical refrigeration, water heating, and cooking cheaper than it was under existing rates. If all customers could be billed for a service charge covering customer costs and perhaps a portion of demand costs (the remainder of which would be covered in the first or second block of the commodity rate), any additional charge would cover merely the energy or output costs, which would be quite low. Customers would then be induced to use more gas and electricity, and there would be a better utilization of plant and distribution facilities.[49]

In the last decade, arguments for such rates have been widely presented. The reports on rates and rate structures of several committees of the American Gas Association, the National Electric Light Association (and its successor, the Edison Electric Institute), contain analyses supporting such fixed-charge rates, and the committees have strongly recommended their adoption by member companies. Committees reporting on rates to the National Association of Railroad and Utilities Commissioners have often favored these types of rates, as have the regulatory commissions of many states.

Recent Trends in Rate Usage

In the manufactured-gas industry, rates embodying some kind of service charge came into widespread use during the period from 1925 to 1933, and, as is shown in the table below, increased in proportion to the total schedules in effect for all classes of service. The total number of manufactured-gas schedules in effect increased by 132.4 per cent from 1925 to 1933; in the same period the number of service-charge rates of all types rose 300.3 per cent. Indeed one service-charge type—

49 For a more detailed consideration of this argument see Chapter III, *infra.*

TABLE 1

FREQUENCY AND PERCENTAGE DISTRIBUTION OF TYPES OF MANUFACTURED-GAS RATES FOR ALL SERVICE CLASSIFICATIONS, 1925 AND 1933 [(a)]

Type of Rate [(b)]	Number		Percentage	
	1925	1933	1925	1933
Service-charge rates				
Fixed service charge	114	336	10.2	13.0
Initial-charge rate [(c)]	163	741	14.5	28.5
Three-part demand rate	21	116	1.9	4.4
Total service-charge rates	298	1193	26.6	45.9
Non-service-charge rates				
Straight-line and step rates	275	240	24.5	9.2
Block rate	516	708	46.0	27.1
Two-part demand rate	27	383	2.4	14.7
Wright demand rate	6	80	0.5	3.0
Miscellaneous	0	3	0.0	0.1
Total non-service-charge rates	824	1414	73.4	54.1
Total all rates	1122	2607	100.0	100.0

(a) Sources: A. G. A., *The Trend of Gas Rate Structures in the U. S. During Recent Years* (New York, 1933), Tables 1 and 2; A. G. A., *The Trend of Gas Rate Structures in the U. S., 1925-1930* (New York, 1930), Tables 1 and 2.

(b) Includes all customer classifications; does not include prepayment rates (196 in 1933), nor liquified petroleum rates (131 in 1933).

(c) The initial charge is more accurately designated as a "block rate with a small initial block constituting a minimum bill"; see p. 14, *supra.*

the initial-charge rate—had become the most widely used of all the various types of rates in effect in 1933, even surpassing the block rate which was rapidly waning in favor. Although the adoption of service-charge rates has not gone far in the natural gas industry, the percentage distribution of rate types for 1933 is strikingly similar to that for manufactured-gas rates in 1925.[50]

The trend toward the use of service-charge rates is even more marked in the field of domestic and general service rates than it is when all classes of service are considered together,

50 A. G. A., *The Trend of Gas Rate Structures in the U. S. During Recent Years* (1933), pp. 24, 25.

as shown in Table 2 below. For similar rates in the natural gas industry, the trend toward the service charge had not gone as far by 1933 as it had in the manufactured gas industry. In 1930, rates of the service-charge type constituted 27.8 per cent

TABLE 2

SERVICE CHARGE PROVISIONS FOR DOMESTIC OR GENERAL SERVICE CLASSIFICATIONS IN MANUFACTURED-GAS RATES, 1925 AND 1933 [a]

Type of Rate [b]	Number of Rates in Effect		Percentage Distribution	
	1925	1933	1925	1933
Service-charge rates				
Initial charge, plus straight-line rate	11	45	1.1	3.6
Initial charge, plus block rate	134	422	14.1	34.2
Service charge, plus straight-line rate	36	49	3.8	4.0
Service charge, plus block rate	61	109	6.4	8.8
Service charge, plus step rate	3	2	0.3	0.2
Three-part rate	1	9	0.1	0.7
Total service-charge rate	246	636	25.8	51.5
Non-service-charge rates				
Straight-line rate	231	76	24.2	6.2
Step rate	14	6	1.5	0.5
Block rate	459	466	48.2	37.7
Two-part demand rate	3	18	0.3	1.5
Wright demand rate	0	32	0.0	2.6
Total non-service-charge rates	707	598	74.2	48.5
Total all rates	953	1234	100.0	100.0

[a] Source: A. G. A., *The Trend of Gas Rate Structures in the U. S. During Recent Years*, Table 4.

[b] These include all optional rates for domestic or general service, excluding special water-heating or house-heating rates; prepayment rates are omitted.

of all natural gas rates for domestic or general service; by 1933 this percentage had risen to 41.3. The initial charge was again of importance, for it entered 16.1 per cent of all domestic natural gas rates in 1930, and 28.2 per cent in 1933.[51]

51 *Ibid.*, p. 27.

In rate schedules which lack a service, customer, or initial charge, the minimum charge is often used as a method of recovering some customer costs. The data in Table 3 show that a small and decreasing number of rate schedules for manufactured gas lack either minimum or service charges. This decline is traceable to the growing use of initial- and service-charge rates. Though by 1933 plain minimum-charge rates had diminished both absolutely and relatively, they were still more numerous than any other type of rate.

TABLE 3

MINIMUM AND SERVICE CHARGES IN DOMESTIC AND GENERAL RATES FOR MANUFACTURED GAS, 1925 AND 1933 [a]

Type of Rate [b]	Number of Rate Schedules		Percentage Distribution	
	1925	1933	1925	1933
No minimum or service charge	94	56	9.9	4.6
Plain minimum charge	613	542	64.3	43.9
Initial charge [c]	145	467	15.2	37.8
Service charge	101	169	10.6	13.7
Total	953	1234	100.0	100.0

[a] Source: A. G. A., *The Trend of Gas Rate Structures in the U. S. During Recent Years,* Tables 5, 7; A. G. A., *The Trend of Gas Rate Structures in the U. S., 1925-1930,* Table 20.

[b] Prepayment schedules not included.

[c] The initial charge rate may also be classified as a minimum charge, but here it is distinguished from the plain minimum charge and the plain service charge for it possesses characteristics of both.

Complete data showing trends in the use of types of electrical rates for all classes of service, or the domestic class alone, are not available. The few studies extant do not show clearly how frequently service, initial, customer, and minimum charges are used. Some information may be gleaned from a survey of electrical rates in effect in 1921 and 1929, in cities of 20,000 population and over.[52] In 1921 there were 438 schedules for

52 Walter E. Caine, *Uniform Rate Areas in the Electric Light and Power Industry* (Thesis presented for the M. B. A. degree at Northwestern Uni-

domestic lighting or residence service (excluding special appliance rates), of which 46 (or 10.5 per cent of the total) had service charges, and 38 (or 8.7 per cent of the total) included room or area charges. In 1929, there were 622 rate schedules for residence and lighting service. Of these, 135 (or 21.7 per cent) were service-charge rates and 204 (or 32.8 per cent) were room or area rates.

An analysis by L. R. Nash of rates in effect in 1932 supplies further information on electric rate usage.[53] Nash's data cannot be compared with those reported by Caine; the two studies do not select and classify rates in the same way. According to Nash, at least 14 per cent of the general residential rates in effect in 1932 contained customer or initial charges in combination with straight-line, step, and block rates, and 37 per cent were room and area rates. Judging from these analyses, the initial charge had not come by 1932 to be as common in residential electric rates as it was in domestic gas rates. On the other hand, area rates were more frequently used for electrical service than for gas service. It is evident, however, that in both industries, there has been a growing use of service and customer charges, and these have come to be an important feature of rate structures.

versity School of Commerce, Chicago, May 1931); Appendix A, pp. 11, 12, 17. This analysis is based on the rate-books of the National Electric Light Association for 1921 and 1929.

53 L. R. Nash, *P. U. Rate Structures*, pp. 162-163.

CHAPTER III

THE CASE FOR THE SERVICE CHARGE [1]

Removal of Discrimination Between Customers

The basic argument for the service charge is that it remedies a vital defect of simple forms of meter rates under which the consumers of small amounts of service do not pay enough to cover the costs of serving them. Under a pure meter rate, a consumer pays nothing if he takes no energy or gas. Yet the company must incur customer and demand costs, in order to maintain the connection to the customer, regardless of what he consumes. If the losses on the small customer are not to be borne by the company as a cut in the fair return, they must be shifted to the large customer in the form of a higher charge per unit than would be necessary otherwise. But it is often difficult to shift the burden of customer and demand costs to the larger customers, for the rates for large consumption are limited by the competition of alternative services. The service charge is said to be the most practicable and equitable solution to the problem of the small consumer.

If the small consumer were the exception, the problem of recovering from him his full share of costs would not be serious. But, according to the testimony of many people in public and private office, the number of small customers who do not pay their way usually constitutes a large proportion of all those served. It is often said of the gas industry, almost as an axiom, that where a flat meter rate is in effect about one-

1 The following arguments appear throughout the general literature on the making of rate schedules; hence it is not necessary to cite all the places in which they occur. Representative sources are: Clyde L. Seavey, "The Disputed Service Charge," *P. U. Fortnightly*, VI (1930), 591-597; Committee on Rate Fundamentals, American Gas Association, *Better Forms of Rates for Gas* (1927); H. E. Eisenmenger, *Central Station Rates*; L. R. Nash, *P. U. Rate Structures*; the writings of C. S. Reed, especially a series of articles appearing in the *A. G. A. Monthly*, VIII (1926), 19-22, 717-720; IX (1927), 589-590; X (1928), 275-277.

third of all customers are served at a loss; they fail to pay the operating expenses chargeable against them, to say nothing of the return on the investment required to serve them.[2] Another third of the consumers pay the operating costs incurred to serve them, but only a part of the return on the property devoted to their use. The remaining third pay not only their full share of expenses and return, but also enough to compensate the company for its losses on the first two-thirds. This generalization seems to be supported by the fragmentary data available. Table 4 summarizes data cited by several witnesses in an important rate case in which the service charge was an issue.[3]

TABLE 4

ESTIMATES OF THE PERCENTAGE OF GAS CUSTOMERS SERVED BELOW COST, AT COST, AND ABOVE COST [(a)]

Witness	Below Cost	At Cost	Above Cost
F. C. Freeman	40%	30%	30%
C. S. Reed	33-1/3	33-1/3	33-1/3
J. G. King	30	40	30
T. C. O'Hare	30	35	35
J. D. Shattuck	40+	30-35	20-30
A. I. Phillips	40+	30-40	20-30
H. L. Doherty	66-2/3		33-1/3
L. O. Ripley [(b)]	50		50

[(a)] Though there was no standard definition of "below cost," "at cost," or "above cost," these terms were used approximately as follows: "below cost" meant below cost of operation and maintenance, with no return on investment; "at cost" meant full cost of operation and maintenance, but with no, or only a partial, return on investment; "above cost" meant all operating costs and return, plus enough to balance all or part of the deficits on the two preceding groups.

[(b)] Ripley's estimate squares with a later estimate of Frank A. Newton, who stated in 1929 that 50% of the residence customers of the great majority of gas utility companies of the country were being served at a loss, or at rates which produced no return on the property devoted to their use; "The Rate-making Function of a Utility Management," *Journal of Land and Public Utility Economics,* V (1929), 130.

2 "Report of the Committee on Public Utility Rates," *N. A. R. U. C. Proceedings,* XXXIX (1927), 442-443; L. R. Nash, *Public Utility Rate Structures* (1933), p. 76, W. G. Raymond, *The Public and its Utilities* (1925), p. 123.

3 United States Light and Heat Corp. *v.* The Niagara Falls Gas and Electric Light Co., et al., *Transcript of Record,* 47 Fed. (2d) 567 (C. C. A. 2d, 1931).

The estimates cited refer to companies serving at flat meter rates, throughout the country, as of the years 1925 or 1926. The data given in Table 5 have been gleaned from the testimony of witnesses in the same case. They pertain to specific territories of which the witnesses had first-hand knowledge and in which flat meter rates prevailed, except as noted. In only eight of the twenty-five instances does the combined percentage of customers served at, or below, cost fall below 66⅔ per cent. In eleven instances, this combined percentage exceeds 75 per cent. In one instance, Denver, the customers served at and below cost constituted 91 per cent of all customers under the flat rate; this probably was the result of the transfer of larger customers to the more favorable three-part optional rate.

As might be expected, the proportion of gas sold to customers served at a loss is much less than the proportion of customers as such served at a loss. Comparative data for a few localities are tabulated below.

Locality	Percentage of customers served:			Percentage of gas sold:		
	Below Cost	At Cost	Above Cost	Below Cost	At Cost	Above Cost
Salt Lake City	52%	32%	16%	16%	31.4%	52.6%
Dallas	54	31	15	15	27.5	57.5
Fort Worth	50	32.5	17.5	16	31.5	52.5

The striking fact as to the three cities is that approximately 50 per cent of the customers were served at an out-of-pocket loss, and took only 15 per cent of the gas sold; the above-cost customers, amounting to approximately 16 per cent of the total, took between 52.5 per cent and 57.5 per cent of all gas sold.

This evidence illustrates the contentions of those who have argued for the service charge in the last decade or so. It is difficult to know how much weight to give to this testimony. The data are based upon estimates in some instances and upon fairly careful calculations in others; hence they are not strictly

TABLE 5

ESTIMATES RELATING TO SPECIFIC TERRITORIES AS TO THE PERCENTAGE OF GAS CUSTOMERS SERVED BELOW COST, AT COST, OR ABOVE COST (ABOUT 1926) [a]

Witness [b]	Territory	Below Cost		At Cost	Above Cost
O. H. Fogg	New York City [c]	35–40%		20%	40–45%
E. H. Merritt	Niagara Falls	70		17	13
H. C. Deffenbaugh	Rochester, N. Y.	31.5		25.5	43
A. B. Roberts	Cincinnati	54		26	20
F. A. Newton	Michigan & Ill. [d]	20–25		25–35	40–50
T. N. Purcell	Chicago [f]	33		47	20
G. C. Mathews	Wisconsin		50+		
Ewald Haase	Milwaukee [f]	35		26	39
C. W. Green	Kansas City, Kan.		74		26
G. J. Swan	Topeka [e]		64.5		35.5
McG. Smith	Tennessee Gas Co.		50		50
McG. Smith	Nashville		60+		40
W. A. Dunkley	Memphis [f]	46		24	30
F. McDonnell	Montgomery, Ala.	47		24	29
A. B. McCord	New Orleans [g]	33		25	42
E. C. Connor	Texas	50		10–20	30–40
E. C. Connor	Dallas	54		31	15
E. C. Connor	Fort Worth	50		32.5	17.5
J. B. Porter	Tulsa [f]	68		6	26
G. H. Horning	Salt Lake City	52		32	16
V. L. Board	Denver [h]	50		41	9
H. J. Flagg	Spokane	68.5		17.5	14
H. J. Flagg	Tacoma	31		50	19
H. J. Flagg	Seattle	50		30	20
J. T. Ryan	Pacific G. & E. Co.	15		58	27

[a] Source: United States Light and Heat Corp. *v.* The Niagara Falls Gas and Electric Light Co., et al., *Transcript of Record,* U. S. Circuit Court of Appeals, Second District, 1931. Rates generally are flat meter rates, and often include minimum charges; exceptions are noted.

[b] All testified on behalf of the Niagara Falls Gas and Electric Company except E. H. Merritt (Manager of the Niagara Falls Co.), whose testimony was offered by the United States Light and Heat Corporation. Of the twenty-eight witnesses whose estimates appear in Tables 4 and 5, three (Mathews, Smith, and McDonnell) were affiliated with regulatory commissions, and one (Flagg) was a former commission engineer. The remaining twenty-four were utility executives or employees, or consulting engineers. All testimony of these witnesses was offered in proof of the necessity of the service charge for the Niagara Falls Company; the evidence tendered in opposition to it was meager.

[c] Territory of the Consolidated Gas Co. of New York.

[d] Based on nine towns in Michigan, three in Illinois, and one in Indiana.

[e] Service charge plus flat rate.

[f] Block rate.

[g] Step rate.

[h] Flat meter rate with optional three-part rate.

comparable. The testimony given makes it clear that the cost calculations differ as to what constitutes "at cost," or "below cost." Again, the height of the meter rate in effect in each instance affects the proportion of customers in each cost group. Yet the data presented in the Niagara Falls case point to the conclusion that a large portion of gas customers served under simple flat meter or block rates, such as were common in 1925 or 1926, were served at an out-of-pocket loss, or failed to pay a fair return upon the property devoted to their use. And it is commonly thought in the gas industry that this condition has not changed since then. In 1933 the A. G. A. Rate Structure Committee declared that low rates could not be offered while 65 to 80 per cent of gas customers were served at a loss.[4]

A question naturally arises: what methods are used by the various witnesses to ascertain the amount of cost to be borne by each customer, and hence to determine the percentage of customers served at a loss? What constitutes a proper customer cost is discussed in detail in Chap. IV. It is enough to say here that gas men usually have taken a broad view of the costs to be borne by customers independently of their consumption. They often allocate a large portion of the investment in disbution mains to customers, as well as meters and service pipes, and thus arrive at a relatively large figure for customer costs. If customer costs were restricted solely to those directly and clearly traceable to customers as such, the amount would be much smaller. Conclusions as to what proportion of customers is served at a loss would be altered if customer cost were not taken to include costs related to distribution mains. Though the adoption of a narrower view of customer costs would not reduce the sum of total costs of service, it would shift the line of demarcation between those customers served at cost or more, and those served at a loss, so as to diminish considerably the number of the latter.

4 "Report of the Rate Structure Committee to the National Directing Committee of Executives," *A. G. A. Rate Structure Committee Report*, 1933, p. 38.

Deficit Customers in Electricity Service

In domestic electric service, as in the gas industry, it is commonly said that a large proportion of customers is served at a loss under elementary types of rates. W. A. Jones, in his presidential address to the National Electric Light Association in 1931, estimated that 6,000,000 domestic electric customers in the United States were served at a loss, or without profit.[5] This was slightly less than 30 per cent of all domestic customers on record at the end of 1930. Estimates for specific areas place the proportion of unprofitable customers at even higher figures. Nash, writing in 1933, declared that more than 50 per cent of all domestic electric customers in typical cities of moderate size were unprofitable, and of these one-half were served at an out-of-pocket loss.[6] The unprofitable customers used less than 20 kw.-hr. per month; those who took less than 10 kw.-hr. per month were deficit customers.

Newton has cited the case of an electrical utility serving a large uniform-rate territory (almost 500 communities with an average population of 4,400 people, within an area of 48,000 square miles), whose rates were as low as 1½c. per kw.-hr. for cooking purposes, and whose average revenue per kw.-hr. from residential service was lower than the national average.[7] Despite those stimulating factors, 73 per cent of customers took less than 50 kw.-hr. per month, and averaged less than 21 kw.-hr. per month; the remaining 27 per cent had an average monthly consumption of 177 kw.-hr.

The experience in any given instance may vary widely from the "normal" condition. The proportion of customers served above, at, or below cost depends upon several factors, such as rate levels, the extent of appliance ownership, the economic status of the population, and the like. It also depends upon the method of determining the customer and other costs of serving

5 *N. E. L. A. Proceedings*, LXXXVIII (1931), 3.

6 L. R. Nash, *P. U. Rate Structures*, pp. 157-158.

7 F. A. Newton, "This Business of Rates," *Electrical World*, C (1933), 770.

each consumer. If demand costs are averaged among customers without regard to actual demands, the cost of serving small customers may be overstated, and so also the losses on them. Nevertheless, it can hardly be disputed that some customers do not pay enough to cover costs directly allocable to them.

These conditions support the argument for the service charge as a means of removing discrimination between customers. It is worth while to summarize this argument briefly. The losses on small consumers under a flat or block meter rate must be recovered from large consumers in order to get a fair return. But if this is done, the rate structure then discriminates against the large customer. Such discrimination is said to be all the more unjust because the convenience user is usually well-to-do and can easily pay his full share of costs.[8] Hence, it is argued, the service charge equitably remedies the discriminatory aspects of rates which do not recognize customer (and sometimes demand) costs.

Assurance of Revenues

Another reason for the service charge is that it acts as a revenue guarantor, for it protects the company from a drop in revenues should the amount of service taken by customers decline. Several possible causes of a drop in consumption, and revenues, are:

(a) improvements in the efficiency of gas or electrical lamps or appliances;[9]

8 The factual basis for this claim is examined later; see *infra*, pp. 99 ff.

9 An outstanding example was the introduction of tungsten-filament lamps after 1907. Some utility operators believe that if and when new types of lighting equipment, such as neon and other luminous vapor tubes, are perfected in the future, the diminution of energy consumption might again cause a drop in revenues, and create a serious problem. Though the loss of revenue might be temporary (for it might be counterbalanced by lavish use of the cheaper lighting), it would be embarrassing. Service and demand charges might well be used to maintain revenues. See Nash, *P. U. Rate Structures*, p. 73; W. Harrison, "Probable Applications of Gaseous-tube Lighting," *Electrical World*, C (1932), 694.

(b) the inability of a utility company to retain business in the face of competition of alternative services;

(c) a decline in the economic functions of the home, such as cooking, laundering, and heating, which call for the use of gas or electricity;

(d) the replacement of manufactured gas with cheaper natural gas; this reduces the cost to the consumer of a given amount of heating or cooking, effects a reduction in gas revenues, and also itensifies competition with electricity.[10]

If a service charge were not introduced, a general increase in the quantity rates would be necessary. Such a step would further discourage consumption. The service charge yields a substantial and steady block of revenues. This "back-log" would not be touched by a decline in the quantity of consumption, though it would be affected by a drop in the number of customers served. Certainly a forceful reason for the adoption of service-charge rates by gas companies has been a decline in the average revenues received from customers.[11]

The service charge does not prevent the customer from benefiting by technical advances. Under a flat meter rate, he benefits by a cut in his bill proportionate to the increase in appliance efficiency or service economy. If a service-charge rate is in effect, the consumer benefits by getting more light or heat or cold without using more kw.-hr. or cu. ft., and without paying a higher bill.

10 Nash, *op. cit.*, p. 89. Even a "therm" rate would not give complete protection unless no reduction were made in the rate of charge per therm. The therm rate is one based on British thermal units (B. T. U.) rather than cubic feet of gas.

11 See, for instance, *Re* Customers of Boston Consolidated Gas Co. (Mass. D. P. U.), P. U. R. 1929 E 9. An important reason for the authorization of a service charge in this case was that average revenue per customer had fallen from $45 a year in 1922 to $37 a year in 1928, despite increases in capital investment and taxes that were not offset by economies in operation.

Stabilization of Revenues

Another advantage of the service charge is that it smooths the seasonal variations in monthly bills. The first effect of a new service charge is to increase the bills of many customers in summer months when consumption is low. But these additions are offset by reduction in bills for winter months when consumption is larger. Thus the service-charge rate moderates the seasonal variation in monthly bills. Customers often complain that they cannot understand why their bills are larger during winter months than they are during the summer months. The service charge is said to reduce such complaints and to improve public relations. But the cessation of complaints by these customers is offset, to some extent, by complaints of those customers who expect to get low bills in summer months when they use little service, or when they go away on vacations.

An important cause of instability in gas and electric revenues is the cyclical fluctuation of business. Economic deflation may be reflected in a loss of domestic customers, or in smaller sales to many customers. The use of a service charge would not stop the loss of customers, but it would limit the drop in revenue from those of the remaining customers who decreased their consumption. Little attention has been paid to this possible advantage of the service charge, largely because during the past depression the decline in the consumption of electricity for ordinary domestic lighting and small appliances was more than offset, on the whole, by a rapid increase in consumption for domestic refrigeration.[12]

12 The service charge would be of little use in checking a decline in revenues from large power or lighting customers. It represents only a small part of the cost of serving such customers and is far overshadowed by the demand charge. Even the latter, however, has not always been a guaranty of steady revenues. In a period of depression industrial customers usually ask for, and often receive, reductions in demand charges in order to reduce their overhead costs. Moreover, the maximum demands of power customers often decline as their plants become idle, and thus further reduce demand revenues. See, L. G. Cannon, "The Wholesale Electric Business in the Depression," *Journal of Land and Public Utility Economics*, IX (1933),

Though utility companies might gain by charging rigid prices to restrict the losses of revenues in a time of depression, it is not certain that such prices would serve the interests of society. One of the greatest difficulties in the control of business cycles is the dislocation in the price system which is caused by the failure of certain prices to move in harmony with other more flexible prices. This is not to say that prices for utility service are alone at fault: time may show that flexible prices must be made rigid, rather than rigid prices flexible. At the present writing, however, there is grave doubt that greater rigidity in the general price structure is at all desirable.[13]

The Use of Service Charges in the Design of Promotional Rates

The service charge is not merely a device for equitably distributing costs of service among customers. If it were only that, with no effect upon company revenues, few companies would be likely to introduce it in any form.[14] The fact is that

1-9; and "The Future of the Electric Utility Industry: Fated or Planned?," *Journal of Land and Public Utility Economics*, X (1934), 150-151.

13 This barely opens the door to even more complex problems. It seems certain that utility representatives would oppose suggestions to make utility rates more flexible. L. G. Cannon, for instance, has objected to a scheme which would offset high rates in times of good business with low rates in bad times. Such a plan, he says, would tend to destroy the concept of rate-making based on a fixed valuation and a fixed return, and would tend "to subject the utilities to all the weakening processes of the business cycle." ("The Wholesale Electric Business in the Depression," *Journal of Land and Public Utility Economics*, IX [1933], 8.) But, as J. K. Galbraith remarks in an illuminating discussion of rigid monopoly prices, "No longer is public policy toward monopoly power concerned merely with fair prices. . . . It is also concerned with the problem of *harmonious* prices and *harmonious* price changes; the problem of prices which do not intensify and prolong fluctuations in the economy." (See, "Monopoly Power and Price Rigidities," *Quarterly Journal of Economics*, L [1936], 474.)

14 It has been alleged that if utility companies could collect a rate as high as they wished, they would have no interest whatever in a separate service charge, even though the removal of discrimination would be in the interest of customers. (C. O. Ruggles, "The Significance of Minimum and Service Charges in Rate-making for Public Utilities," *Journal of Political Economy*,

they favor the service charge because it facilitates the loading of customer and demand costs in the first portion of the rate, thus relieving the remaining blocks of a large part of this burden. The follow-on rates may then be quoted at a level more closely approximating commodity costs than before. Such a rate is commonly called a "promotional" or "inducement" rate because the low follow-on charges are said to promote or induce a greater use of service.[15]

Few writers who favor the service charge go so far as to insist that it is indispensable to a promotional rate. Many, however, insist that it is the most feasible means of formulating such a rate schedule.[16] The reasons for the use of promotional rates may be reviewed briefly. So long as the domestic load was chiefly electric lighting or gas cooking, few attempts were made to stimulate domestic consumption because the prospects for larger sales were visibly limited. But since the early and middle 1920's, the spreading use of new and improved types

XXXII [1924], 67; Kinch *v.* Concord Light and Power Co., 6 N. H. P. S. C. R. 381 [1921].) This would be true only if companies were actually limited to a fair return.

15 The literature on promotional rates has developed chiefly in the last decade, though the fundamental idea of the promotional rate goes back to early papers on gas and electrical rates, such as those by Hopkinson, Wright, Greene, and Doherty. The use of the term "promotional rate" is of more recent origin, dating back to about 1925. No attempt will be made herein to give a complete list of sources consulted, or relied upon. Almost every discussion of gas and electrical rates deals with, or hinges upon, the problem of promoting the use of service in order to increase the consumption by customers and the revenues of utility companies.

16 The following is a representative list of authorities favoring the service charge as a promotional device: Watkins, *op. cit.*, p. 143; C. S. Reed, "Make the Unprofitable Customer Profitable," *A. G. A. Monthly,* X (1928), and in his other articles cited above; N. A. R. U. C. Committee on Public Utility Rates, "Report," *N.A.R.U.C.Proceedings,* XLI (1929), 330-331; N.E.L.A. Rate Research Committee, *Electric Rates and their Status in 1929* (1929), p. 3; Nash, *P. U. Rate Structures,* pp. 154, 173, 280; M. L. Cooke, *What Price Electricity for Our Homes?* (1928), p. 17; C. L. Seavey, *loc. cit.,* p. 597; W. A. Jones, "President's Address," *N. E. L. A. Proceedings,* LXXXVIII (1931), 3; *Report of the Rate Structure Committee,* A. G. A., 1933, p. 36.

of electrical and gas appliances has changed the character of domestic consumption, especially in the case of electricity, for which data are shown in the subjoined tabulation. These data

PERCENTAGE DISTRIBUTION OF KILOWATT-HOUR SALES TO RESIDENTIAL CUSTOMERS [a]

Year	Lighting	Major Appliances	Small Appliances	Total Consumption
1920	80%	10%	10%	100%
1926	65	20	15	100
1933	40	45	15	100

[a] Source: "Domestic Appliance Load: 1920-1933," *Edison Electric Institute Bulletin*, II (1934), 1932. The apportionment of consumption between lighting and appliances probably overstates that by appliances, and understates that by lighting. See, W. F. Kennedy, *The Objective Rate Plan* (New York: Columbia University Press, 1937), pp. 2-3.

emphasize the importance of the major appliances (including refrigerators, ranges, and water heaters), which also loom large in the future growth of the domestic load.

One effect of the spread in the use of new appliances was to cause utility managers to reconsider the possibilities of domestic consumption, and to re-examine rate structures. They saw that new rates would lead to the use of more appliances and thus of more gas or electricity. The demand of many (but not all) appliances was comparatively steady, or came at times of the day or season when ordinary domestic demand was slack. Appliance loads usually improved the diversity factor and the load factor of the domestic class of business.[17] Thus they could be taken on with little or no addition to the facilities for serving residential consumers.[18] The amount of gas or electricity actually consumed by appliances cost little to produce or generate. Consequently, appliances could be served at rates which were considerably less than those for electric lighting or gas cooking. The companies gained additional revenues and

17 See Appendix A as to the effect of appliance utilization on the characteristics of domestic loads.

18 Such loads could not be added indefinitely without increasing the investment. But thus far the added loads have fallen well within present capacity. See *infra*, p. 60, as to exceptional cases.

net incomes, which would not have been obtained otherwise. Domestic customers benefited by the use of gas and electricity in larger quantities at a lower cost.

Rates for appliance business are made low not only because the cost of service is low, but also because the business is competitive. Electricity and gas compete with each other, as well as with other goods and services. The rates for a given service have to contend with the cost of using substitute services, even though the various services differ in convenience, adaptability, cleanliness, safety, and prestige.[19] Of course such differences often limit competition. In some uses, one service is far superior to the others at prevailing rate levels. Thus electricity does not now face serious competition from gas for illumination. Likewise the supremacy of electricity for small appliances, such as flatirons, washing machines, vacuum cleaners, and radios, is unchallenged. The result is that rates for electricity can be made very high before people will resort to alternative sources of light, such as gas, kerosene, gasoline, or candles, or stop using small appliances.[20]

In both the gas and electrical industries, it has been urged that forms of rates be adopted that are suitable for getting and

19 For example, for water heating, electricity at one and one-half cents per kw.-hr. is equivalent in price to manufactured gas at $1.50 per thousand cu. ft. of a calorific value of 540 B. T. U. per cu. ft., or $2.50 for a mixture of manufactured and natural gas of a calorific value of 900 B. T. U. per cu. ft. This does not take into account the customer's investment in equipment. (Data from N. R. Gibson, "Rational Distribution of Electric Energy," paper read before the World Power Conference, 1936, p. 6.) Gibson gives a table showing the comparative costs of gas, electricity, and coal for water heating and cooking. For a formula which permits the derivation of the comparative costs of gas and electricity at various prices see, Irving Bussing, *Public Utility Regulation and the So-Called Sliding Scale* (New York: Columbia University Press, 1936), Appendix, pp. 168-169.

20 Nash estimates that 80% of domestic electric service (in kw.-hr.) is subject to more or less direct competition. A customer who takes 600 kw.-hr. per year, applies only about 200 kw.-hr. to lighting and small appliances, and the remaining 400 kw.-hr. are applied in competitive uses. In a home fully electrified, more than 90% of a probable consumption of 6000 kw.-hr. per year competes with other services. Nash, *P. U. Rate Structures*, pp. 153, 322; see also, Bussing, *op. cit.*, p. 169.

retaining business. Service-charge rates especially have been favored by utility operators and rate committees.[21] Other changes in rates do not meet the dual problem of reducing rates for large consumption and at the same time maintaining revenues from small consumers of gas or electricity. A general rate cut might bring success on the competitive front, but would allow the small customers, who had not been paying their way under the old rate, to be served at an even greater loss under the new rate. A reduction in the rates for the first one or two blocks would be even worse. Large customers would not be relieved of their overcharge; small customers would receive lower bills though they would not be likely to take more service. If the reduction in rates were confined to the competitive business alone, the utilities contend that the over-all fair return would be impaired. Nor would a cut in follow-on rates remedy the maladjustment of costs between large and small customers. "It would seem, therefore, that the only solution to the difficulty lies in a readjustment of charges as between customers, through the incorporation into the rate structure of a stand-by (fixed) charge of some nature." [22]

Competitive prices for different sections of the residential market have been stated in two ways. First, as special rates for sharply restricted appliance loads. Sometimes the special rate does not apply unless the appliances are wired on a separate circuit and meter. Second, the competitive rate may be stated as a single rate of the block type, for all domestic customers for all uses. The first method insures that an appliance rate, such as for water heating, is enjoyed by a customer only for the demand and consumption of the specific appliance. But it adds to the costs of service, for the company must install an extra

21 *Report of the Rate Structure Committee,* American Gas Association, 1931, p. 25; *ibid.*, 1933, pp. 36, 37; Alabama *v.* Alabama Public Service Commission, P. U. R. 1925 A 413, Nash, *op. cit.*, p. 179.

22 "Report of the Subcommittee on Presentation of Stand-by Charge Rates," *Report of the Rate Structure Committee,* A. G. A., 1931, pp. 25-26. See also, *ibid.*, 1933, p. 37; F. A. Newton, "This Business of Rates," *Electrical World,* C (1933), 773.

meter and the consumer a separate house-wiring circuit. This drawback is avoided by a single rate for all domestic service, in which proper blocking provides the desired differentiation in pricing the service. The single rate schedule is also in keeping with the movement toward simpler rate structures.

The problem involved in making the single rate is to block the rate so as to distinguish between the uses in which demand is relatively inelastic, and the uses to be developed, in which the economic demand is relatively elastic. The number of blocks to be used, and the height and the width of each block, must depend upon the characteristics of each market. For example, in 1929 the Committee on Public Utility Rates of the National Association of Railroad and Utilities Commissioners recommended that electric rates consist of an initial-block charge plus two follow-on blocks. The rapid development of the appliance load in following years led the Committee to recommend in 1935 that more than two blocks should be used if necessary. The Committee specified that the width of the first two blocks (the first of which was covered by the initial charge) should equal the average consumption for lighting and small appliances. The rates for these blocks should cover fixed costs as well as energy costs. The third block should equal the average use for refrigeration (say from 30 to 60 kw.-hr. per month), and probably would have to be priced at two to three cents per kw.-hr. Similarly, the electric-range load could be reached in an ending block at one to two cents per kw.-hr. The rates and blocks for additional loads could be fixed in like fashion.[23]

Likewise, in the gas business it has been strongly urged that a single rate of the block type, including a service charge, should be used instead of several special appliance rates. The Rate Structure Committee of the American Gas Association,

23 Committee on Public Utility Rates, "Uniformity of Electric Rate Forms and Simplification of Electric Rates," *N. A. R. U. C. Proceedings*, XLVII (1935), 201-202. For another example of proposed blocking see, W. S. Leffler, "A Study into the Prices of Domestic Electric Service," *N. A. R. U. C. Proceedings*, XLI (1929), 461-463.

in its report of 1933, suggested that the first block after the service charge, either direct or indirect, should be based upon the usual requirements for cooking and refrigeration purposes; the second block should be governed by requirements for water heating, and incineration, and the third block should be priced for the house-heating load.[24] Other variations in blocking gas and electric rates are common. Each rate-maker tries to adjust the rate to the conditions of demand and use, actual and potential, that he faces. In so doing, he expresses his own opinion as to the best and most feasible method of recovering customer and demand costs.[25]

The advocates of the inducement rate say that its great advantage is that it recognizes the reciprocal nature of the relationship between rates, consumption, and costs. If follow-on charges are based upon the added costs of new business, they may then be low enough to stimulate consumption and increase output.[26] At first the revenues may be less than revenues from the old rate. After some time, however, the expansion of consumption restores the revenues to their former or even higher levels.[27] Total costs are not increased in proportion to the increased output. As output increases, the costs per unit of additional service decrease, and thus the company is enabled to reduce further the follow-on rates. Reductions in costs follow a more intensive utilization of plant and distribution facilities. If the increase in output requires production on a larger scale,

24 *Report of the Rate Structure Committee*, A. G. A., 1933, Appendix I, "Report of the Subcommittee on Domestic Gas Rates," p. 12.

25 See Chap. VI, *infra*, for a discussion of the relative merits of the several alternatives to the direct service charge.

26 It is assumed here that the changes in rates are not made in order to reduce the revenues to which companies are legally entitled. Undoubtedly, the question of the "fairness" of the general level of domestic rates is involved in the present-day problem of making rates promotional. But this question cannot be discussed here.

27 The length of the period during which revenues are deficient depends upon a variety of factors, such as the extent of the rate cut and the purchasing power of customers. For a discussion of the evidence on the relationship between rates, consumption, and revenue, see Appendix B.

other economies are effected, such as a lower investment per unit of productive capacity. The most enthusiastic advocates of promotional rates speak of a recurring cycle of lower rates, increased use, and lower costs.

From the point of view of the utility companies, the service charge makes it possible to shift to the small customers the burden of the initial cut in rates for those consumers whose potential demand is large. The utility's revenues are thereby protected. But the entire burden of the rate cut cannot always be shifted. Even with a service charge, the follow-on rates would often have to remain comparatively high to maintain revenues. Many companies, therefore, have had to sacrifice some part of their revenues to initiate a promotional rate. The hope is that the loss in revenue will be recovered when consumption responds to the reduced rates. The prospects of such a response are uncertain. Utility companies often have been unable or unwilling to forego a part of their revenues for a period which may last anywhere from a few months to a few years, despite the protection of a service charge.[28]

28 In recent years, several schemes have been developed which limit the loss of revenue from a rate change to an amount predetermined by the company. (For the details of various plans see, H. Zinder and W. E. Caine, "Bargain Sales for Electricity," *P. U. Fortnightly*, XV [1935], 690-701.) A characteristic feature of these schemes is that the customer, to be eligible for lower rates, must have a bill at least equal to that for the corresponding month of the previous year or for some other "base" period. Thus, the revenue of the utility remains undiminished. The customer's total bill is not reduced unless his consumption decreases. The rate per unit of additional service decreases only when the customer increases his consumption beyond his "base" amount. He thus increases his total bill and also the revenues of the company. As to other details, the several plans vary. But they all represent an attempt to increase the consumption of the smaller customers who have not thus far responded to offerings of special appliance rates. Essentially, such rates are intended to make available immediately to the customer who increases his use, that rate which would be possible only if use were increased. The old dilemma—how to increase the volume of use without reducing revenues?—is attacked by making rate cuts provisional upon the increase in volume of consumption. These schemes, of which the Commonwealth and Southern Corporation's "objective rate plan" has been

In summary, the service charge has been favored on the ground that it permits the formulation of a promotional rate, which leads to an increase in the use of service, with benefits to the company and the customers. The small consumer, whose bills are immediately increased by a service charge, is said to gain the opportunity of buying larger quantities of service than he used before, at a much lower rate per unit. This is said to encourage him to enlarge his consumption.[29] The large consumer also gets this opportunity, as well as a reduction in his bill, even though he does not consume more. The utility company gains larger revenues. The community as a whole is said to gain by the centralization of the supply of light, heat, cold, and power, and by the elimination of individual utilization of coal and other fuels which spread a pall of smoke, dust, and dirt over the city.

Is the Service Charge a Promotional Device?

The argument that the service charge makes possible a promotional type of rate is a powerful one. It seems to justify the universal use of service-charge rates, especially if these actually correct discrimination. Only brief reflection is necessary, however, to suggest that a service-charge rate is not necessarily promotional. Forces other than the type of rate schedule influence the amount and character of consumption in any community. The form of the blocking (the height, width, and number, of the rate blocks), as well as the level of the follow-on blocks, may offset consumption more than does the mere presence of a service charge. Non-rate factors must be reckoned with in giving credit for results of rate changes. Among these

most publicized, are an important development in the art of rate-making. See, F. A. Newton, "The Commonwealth and Southern Objective Rate Plan," *Journal of Land and Public Utility Economics*, XI (1935), 117-122; W. F. Kennedy, *The Objective Rate Plan* (New York: Columbia University Press, 1937).

29 "The service charge type of rate is the best rate for the small customer because it is the only type of rate which gives him a chance to grow into a large customer," C. S. Reed, *A. G. A. Monthly*, X (1929), 277.

are: the availability and the cost of appliances; the expense of house-wiring (or re-wiring) to permit the use of more appliances; the efforts of the new business and merchandising departments of the utility company; the purchasing power and the desires of the various consumer groups in the community; and the availability and price of substitute services. Other important factors may appear from time to time. It must be recognized, therefore, that the introduction of a service-charge rate does not insure promotional consequences.

Considered by itself, the service charge does not make a rate promotional. A high bill for a small consumption does not whet the small customer's appetite for more service. This is especially true if follow-on rates are not blocked so as to offer additional service at a price which the small customer can afford to pay. In fact, a high initial charge leading to a bill of $2.50 to $3.00 for 30 kw.-hr. does not encourage the ordinary customer to extend his consumption, even though, unlike many customers, he may be able to spend more for service. Hence, a high service charge may be demotional rather than promotional in effect.[30]

Even where the service charge is moderate, and where the follow-on rates are skilfully blocked, it may be difficult to expand the sale of gas or electricity because of the inability of important groups of customers to buy more service. Consider the case of a gas company serving large numbers of small customers who live in small flats or apartments supplied with heat, hot water, and perhaps refrigeration, by the landlord. A market of this kind may not offer likely prospects for any substantial sale of the major appliances upon which load development largely depends. The justification for the service-charge rate in this case must rest on the ground that it removes discrimination against larger customers, if that can be shown to prevail.

It must be conceded that, *ceteris paribus,* a service-charge rate is more promotional in effect than one which less clearly

30 See, John Bauer, "If Power Cost Nothing to Produce—What Would Be the Cost to the User?," *P. U. Fortnightly,* X (1932), 381.

differentiates in the price per unit for various quantities of use. This does not mean that a direct service charge is the only type of rate which may be used to formulate a promotional rate. Nor is it the only kind of rate that assures certain and stable revenues. These advantages attend any system of rates that includes charges other than those for mere quantity of use. Rates including demand, initial, or minimum charges, often are promotional in effect, and often avoid the demotional influence of a naked fixed charge. It is hard to make the service charge as large, and hence as promotional, as other forms of other-than-use charges, because small customers bitterly and effectively oppose a large undisguised fixed charge. On the other hand, a small service charge that is not opposed by customers may lose much of the advantage of promotionality. This is the dilemma of the rate-maker who uses the service charge. He must face the fact that the use of the service-charge rate is not justified if the tasks that it is supposed to do are done easily and efficiently by other types of rates that avoid its drawbacks.[31]

Some danger attends the use of promotional rates when new business is taken at rates based upon the additional costs immediately occasioned by it. If consumption and demand increase beyond the extent to which they can be served by existing spare capacity, the new rates may not be sufficiently high to cover the cost of the new system demands, particularly if it is necessary to add facilities for distribution and production.[32] Where such poorly designed rates have been in effect for some time, and have led to the installation of expensive domestic appliances, customers have come to believe that they have established a "right" to receive service at the existing rate.[33] This danger

31 See Chapter VI for a discussion of alternatives to the direct service charge.

32 Nash believes that a policy of not covering full demand costs may be justifiable temporarily, but certainly not permanently; *P. U. Rate Structures*, p. 87.

33 "Report of the Rate Research Committee," *N. E. L. A. Proceedings*, LXXV (1921), 38.

is not necessarily inherent in service-charge rates. But it may arise if the follow-on rates are incremental rates which fall short of covering the long-run incremental costs of service.[34] For example, an increase in the electric range load in a highly industrialized Canadian city of 30,000 people made it necessary to increase the capacity required to meet the peak demand. In this instance, the simple schedule of a customer-charge plus block-energy rate was too promotional. It became necessary to institute demand charges in order to shift costs where they belonged.[35] These dangers are known to the utility industry.[36] Yet they may be, and often are, easily overlooked when rates are designed to get competitive business.[37]

The Service Charge as an Aid to Simplicity in Rate Form

Rate structures for domestic gas and electric service long consisted of a conglomeration of rates inherited from the earlier days of simple domestic loads, plus newer class rates designed to tap special loads. Though this is now less true than it once was, the introduction of a variety of promotional and optional rates has frequently increased the confusion in rate structures. All too often the customer has been confronted by one or more rates so complex in form that he has found them difficult to

34 The long-run incremental costs may not be *full* pro-rata costs, but may be nearer to them than short-run incremental costs. *Cf.* p. 93, *infra.*

35 H. D. Rothwell, "Investigation of Domestic Demand," *Bulletin of Ontario Hydro-Electric Power Commission,* XVI (1929), 36-47. In this town the customary hour for dinner was at noon, and the cooking demand therefore created a peak when added to the normal morning load.

36 "Rates made with the primary purpose of business promotion must especially give appropriate weight to cost analysis under the conditions which will prevail after the estimated load has been acquired," *Report of the Rate Structure Committee,* A. G. A., 1933, p. 3; see also, *ibid.,* 1934, p. 6.

37 ". . . I am sure that many of these promotional rates have been designed to secure business in the face of keen competition, and, as designed, are unsound and dangerous . . ."; W. A. Jones, "President's Address," *N. E. L. A. Proceedings,* XXCVIII (1931), 3.

understand and apply. Hence, within the last few years, there has arisen a strong demand for the simplification of rate practices by wiping out all inconsistent and superfluous schedules, and by recasting the remaining schedule in a simple form.[38] If this were done, customers would more readily understand the rates applying to them. Each customer could easily compute his bill himself and compare it with those elsewhere.[39]

The movement toward simpler types of rate structures is now well under way. Though no single rate type is replacing others, service-charge rates, especially of the initial-charge type, are being favored because of their simplicity and flexibility. Usually the service charge is a part of a block meter rate which is easily explained and applied. The service charge, therefore, is of importance because of its simplicity, in comparison with older forms of rates, and because often it may be made to replace a variety of confusing class rates.

Improvement of Public Relations

Out of these advantages of the service charge, there is said to flow another—the improvement of the relations between the utility company and its patrons. There is little doubt that the elimination of discrimination, the reduction of seasonal variations in monthly bills, and the adoption of lower follow-on rates would tend to reduce criticism of utilities by the public. The specific benefits of better public relations are: fewer complaints about high bills; cheaper administration of utility operations, particularly of the utilization, commercial, and new busi-

38 See, for instance, Paul M. Downing, "Rate Simplification as an Industry Need," *N. E. L. A. Proceedings*, LXXXIX (1932), 23-25; Committee on Public Utility Rates, "Uniformity of Rate Forms and Simplification of Electric Rates," *N. A. R. U. C. Proceedings*, XLVII (1935), 195-198; Alex Dow, "Discussion of Rate Research Committee Report," *N. E. L. A. Proceedings*, XXCI (1924), 77.

39 The Committee on Public Utility Rates of the N. A. R. U. C. asserts that a standard form of rate consisting of an indirect service charge (i. e., initial charge) plus a block energy rate, would facilitate valid rate comparisons, and would prevent many illogical comparisons. *Ibid.*, p. 202.

ness departments; less rate litigation; and an easing of the tasks of regulatory agencies.

Unfortunately, public relations are not always improved by service charges. A great drawback to the service charge, which has thus far blocked its widespread adoption, is that most customers object to it. They have not been convinced by education or propaganda that they are benefited by it. Experience shows that in many instances the immediate effect of a service-charge rate is to harm public relations rather than to improve them. Apparently, much depends upon the circumstances of the rate change. For, in some cases, public relations ultimately have been improved by a moderate service charge that has been skilfully introduced, particularly at the time of a rate reduction.[40]

Conclusions

The case for the service charge rests largely upon two arguments: that it wipes out discrimination, and that it leads to promotional rates. The latter argument has been shown to have a great deal of force. But it also has been shown that the promotional effects of service-charge rates are largely relative, and are not unlimited. Further light is cast upon this aspect in the discussion of the effect of the service charge upon various groups of domestic customers which appears in Chapter V. It remains, therefore, to consider the soundness of the first argument, namely, that the service charge removes discrimination. This calls for a detailed examination of the costs that are said to be the basis of service charges. That is the task undertaken in the next chapter.

40 The causes and consequences of the opposition to the service charge are discussed in Chapters V and VI.

CHAPTER IV

WHAT COSTS SHOULD ENTER THE SERVICE CHARGE

A large part of the case for the service charge lies in the argument that, in its absence, many customers (notably, the small ones) get service below cost. The immediate question is then, what costs should be covered by a service charge? Discussion of this question is made difficult by the fact that there is a wide difference of opinion as to what the substance or content of the service charge should be. Chairman Maltbie, speaking for the New York Public Service Commission has said, " . . . there is a great variety of opinion and little agreement as to what items should be considered in computing a proper service charge, and there is ample room for fantastic assumptions." [1] One student of gas rates found the confusion so great that she concluded, " . . . when the term 'service charge' is employed it is impossible to know what costs are covered by the charge." [2] This lack of agreement as to the content of the service charge has worked against intelligent discussion of the subject, and instead has added to the confusion in public utility rate-making.

A study of rate discussions reveals three main ideas of what the service charge should be, namely: (1) a charge based solely upon demand costs; (2) a charge based upon demand and customer costs; (3) a charge covering solely customer costs. The first may be dropped, for the term service charge is seldom used as a synonym for a demand charge, which is purely a function of the customer's demand for service, and the demand charge alone is beyond the intended scope of this treatise. It may be pointed out, however, that when customer costs are small in proportion to total costs, as in the case of

1 *In re* Electric Companies, P. U. R. 1931 C 337, 346. See also Alabama Public Service Commission *v.* Alabama Power Co., P. U. R. 1925 A 424.

2 Farnsworth, *Rate Structures in the Manufactured Gas Industry*, p. 224.

large industrial consumers, they are usually absorbed in the demand charge.

The second meaning of the service charge, as one covering all or part of customer and demand costs, has been widely held.[3] The term " readiness-to-serve charge " has been used in this broad fashion. Perhaps it expresses the sense of a charge to cover demand and customer costs made necessary by the company's readiness to serve the customer more accurately than does " service charge." [4] Unfortunately, the readiness-to-serve charge has not always been used to mean one thing; sometimes it has meant a customer charge,[5] and at others a demand charge.[6] It is no clearer in meaning than " service charge," and usage sanctions the shorter term.[7]

3 E. Jones and T. C. Bigham, *Principles of Public Utilities*, pp. 341, 342; *Report of the Department of Public Utilities on the Expediency of a Service Charge*, Commonwealth of Massachusetts, House, No. 2, January, 1921; H. C. Spurr, *Guiding Principles of Public Service Regulation* (Rochester, N. Y.: P. U. Reports, Inc., 1926), III, p. 413; W. L. Ransom, *The Service Charge as a Part of the Rate for Gas* (1921), p. 101; W. A. Prendergast, *Public Utilities and the People* (New York: D. Appleton-Century Co., 1933), pp. 219, 223; " Local Chambers Effective in Justifying Demand Changes," editorial, *Electrical World*, CI (1933), 186; Report of the Committee on Public Utility Rates, *N. A. R. U. C. Proceedings*, XLI (1929), 333; C. O. Ruggles, " The Significance of Minimum and Service Charges in Rate-Making for Public Utilities," *Journal of Political Economy*, XXXII (1924), 62-65; A. Parshall, " The Service Charge," *American Gas Journal*, CXIV (1921), 234; N. B. Jacobs, " Need the Service Charge be Uniform to all Domestic Users?," *P. U. Fortnightly*, X (1932), 199; V. M. F. Tallman, *Report of N. Y. Com. Rev. P. S. C. Law*, III, 2275; H. Erickson, " Rates and Rate Making Under the Wisconsin Public Utility Law," *Amer. Water Works Assoc. Proceedings*, XXXIII (1913), 66; C. D. Jackson, " The Trend of Court Decisions and Legislation," *N. E. L. A. Proceedings*, LXXXI (1924), 100.

4 Spurr, *op. cit.*, III, p. 386.

5 For example, Bramhall *v.* Ohio Public Service Co. (Ohio P. S. C.) P. U. R. 1933 E 117.

6 For example, S. S. Wyer, " Notes on Readiness-to-Serve Charges for Public Services," *Case and Comment*, XXII (1916), 1031.

7 C. O. Ruggles, *op. cit.*, p. 58.

The third meaning of service charge, as a charge for customer costs only, is probably the most common.[8] In this sense it is synonymous with the terms " customer charge " and " consumer charge," which are also in frequent use. Another term, " meter charge," has been used occasionally in lieu of service or customer charge, but it is now rare.[9] Moreover, it has been given a special meaning in G. P. Watkins' careful analysis, as a charge that varies with the size of the customer's meter and which may be loaded with some part of demand costs.[10] It will be used in that sense hereafter.

This confusion as to the meaning of service charge, as well as the public antagonism to it, has led to attempts to substitute a new term. In 1931 a subcommittee of the Rate Structure Committee of the American Gas Association proposed that " stand-by charge " be used instead of " service charge." [11] But this term has not met with great success. Though sometimes a " stand-by charge " has been used as a rate based on demand as well as customer elements, it may also refer to a charge for stand-by service.[12] Such service provides a " permanent connection for use in case of emergency or in case the consumer finds his usual source of supply inadequate." [13] In the

8 Federal Power Commission, Electric Rate Survey, *Glossary of Important Power and Rate Terms* (1936), pp. 4, 9; John Bauer, " The Effect of the Service Charge on the ' Small User '," *P. U. Fortnightly*, IX (1932), 269; " Report of Rate Structure Committee," *A. G. A. Proceedings*, XLIII (1931), 179; *ibid.*, XXXVIII (1926), 36.

9 See Commissioner Maltbie's dissenting opinion in, *In re* Stadtlander *v.* New York Edison Co. (N. Y. P. S. C., 1st Dist.), P. U. R. 1915 B 685.

10 G. P. Watkins, *Electrical Rates* (New York: D. Van Nostrand Co., 1921), pp. 70, 97, 120; see pp. 153-7, *infra.*

11 *Report of the Rate Structure Committee* (New York: A. G. A., 1931), Appendix III, " Report of Subcommittee on Presentation of Stand-by Charge Rates," p. 24.

12 See, for instance, *Re* Reedsburg Utility Commission (Wis. R. C.), P. U. R. 1924 C 290; J. C. Stipher, " Charges for Stand-by, Breakdown, or Auxiliary Service," *Journal of the American Water Works Association,* XXII (1930), 942.

13 Electric Rate Survey, *Glossary,* p. 9.

opinion of the writer, therefore, the term stand-by charge should not be used for ordinary electric service, nor, for the sake of harmony, for gas service.

From what has been said above, it is clear that a distinction should be made between the service charge that covers customer costs, and the service charge that includes customer and demand costs. But neither of the two ideas of what the service charge should contain is clear cut, nor are the two mutually exclusive. Those who agree that the service charge should consist of customer costs, often disagree as to what such costs are. Customer costs are not self-defining. One rate-maker may declare that they are merely the expenses of metering the service, and of making out the bill and collecting it from the customer. But another while also thinking of the service charge as a customer charge, says that customer costs include all the costs traceable to the customer as such, as well as a part of distribution expenses and a good share of general expenses. And a third person puts into the service charge the very same items as the second, but calls a part of them demand costs and applies them as such. It is necessary, therefore, to inquire further into the meaning of customer costs.

Customer Costs as those Varying with the Number of Customers

The definition of customer costs that is most widely used is those costs which vary with the number of customers attached to a utility distribution system.[14] These are generally distin-

14 S. E. Doane, "High Efficiency Lamps: their Effect on the Cost of Light to the Central Station," *N. E. L. A. Proceedings*, XXXVIII (1910), 155; "Report of Committee on Public Utility Rates," *N. A. R. P. U. C. Proceedings*, XXVIII (1916), 101; W. G. Vincent, "Analyzing Gas Costs and Sales," *Pacific Coast Gas Association Proceedings*, XI (1916), 267; W. L. Ransom, *The Service Charge as a Part of the Rate for Gas* (New York, 1921), p. 16; W. J. Greene, C. F. Schoonmaker, and C. B. Gorton, "Handling Cost Elements in Fixing Rates," *Electrical World*, LXXX (1922), 1431; N. M. Argabrite, "Fundamental Rate Basis," *Public Service Management*, XLVI (1929), 47; C. H. Dauchy, "Costs in Rate Making," *Electrical World*, XCIII (1929), 733; W. S. Leffler, "A Study into the Prices of

guished from demand and energy costs on the ground that they are independent of the consumer's demand and consumption. However, to say that customer costs are those which vary with the number of customers does not reveal which costs they are. If one takes a broad view, and lumps together as customer costs those costs which vary " roughly " or " largely " with the number of customers, then the definition bears little weight. For almost all costs, if not all, vary in some degree or largely or roughly with the scale of operations of the company, which reflects the number of customers. Such a broad definition of customer costs would not lend itself to the careful analysis of costs, and would always justify any and all customer costs which a company might want to collect from its consumers.

The definition of customer costs as those based on the number of customers must be carefully narrowed if it is to be valid. Customer costs are often defined as those which increase or decrease directly with an increase or decrease in the number of customers. Sometimes it is said that customer costs are those which vary in closer degree with the number of customers than with demand or energy. Though the definition may be carefully stated in this way, there is no great agreement as to precisely which items of expense and investment vary in such a manner. Nevertheless, the tendency is to include a rather broad range of items. This is done on the ground that as each customer is provided with service, he not only requires the installation of an additional meter and service line or pipe, but also a certain capacity in distribution lines or mains and (in the case of electricity) transformers. Certain other expenses are incurred in operating and maintaining this investment, in keeping records of the customer's account, and in billing and collecting.

Domestic Electric Service," *N. A. R. U. C. Proceedings*, XLI (1929), 448; John Bauer, " The Service Charge and its Developments in New York," *National Municipal Review*, XX (1931), 543; " Report of the Rate Structure Committee," *A. G. A. Proceedings*, XIII (1931), 27; and numerous others.

Ordinarily, these items of customer costs are said to be independent of both the demand and the consumption of customers. In the gas industry, however, the distinction is some times drawn merely between costs varying with number of customers, and those varying with output. Consider the position taken by the American Gas Association, summarized as follows.[15] If one imaginary customer, adjacent to a gas plant, had a consumption and a load curve equal to the aggregate consumption and load curve of all existing gas customers, no distribution system other than the large pipe connecting him with the gas plant would be required to serve him. If the gas company served one such customer rather than thousands of customers spread over a wide territory, certain costs would be negligible. These would be the carrying costs on the distribution systems, as well as the costs of meter-reading, bookkeeping, and other commercial expenses, plus the general, miscellaneous, new business, utilization, and other expenses (save production expenses). In the same way, these costs would be much less if the gas company, now serving 100,000 customers, were to serve only 5,000 customers, covering about 5 per cent of the present territory, but with the existing consumption and load factor unchanged. It is asserted that such costs are determined largely by the customers themselves, since a gas utility has no control over the location of their homes. It is concluded, therefore, that these costs "follow much more closely the number and location of the customers than they follow—for example—the number of cubic feet sold." It is not contended that these expenses increase or decrease in mathematically direct proportion to the number of customers, but it is maintained that "they much more properly and equitably follow the number of customers rather than the number of cubic feet or, for that matter, any other commonly used basis such as demand."[16]

15 *Report of the Rate Structure Committee* (New York, A. G. A., 1931), Appendix III, "Report of the Subcommittee on Presentation of Stand-by Charge Rates," pp. 26-27.

16 *Ibid.*, p. 27.

The foregoing argument was suggested by the Subcommittee on Stand-by Charge Rates, as one to be used by gas companies in support of a customer charge determined by a two-part cost analysis which would only yield customer costs and commodity costs. Nevertheless, three- and four-part cost analyses were recognized by the Subcommittee to be necessary for the company's own information. Ordinarily, two-part analyses place all or most of the demand costs in the commodity group, in contrast to the two-part analysis outlined above which places all or most of demand costs in the class of customer costs. Other rate structure committees of the American Gas Association have separated demand and customer costs.[17] Yet the difference has been one of degree. Fundamentally, the view of customer costs held by the several gas rate committees has been the same, namely, a broad one.[18]

Other Definitions of Customer Cost

Many other definitions of customer costs are found in rate literature. Thus, customer costs are said to be those which are approximately equal for all customers regardless of the demand or consumption of the customer.[19] Raymond qualifies this definition by specifying that the customers are classed according to size of service connections. However, there is no unanimity of opinion as to which expenses are largely the same for all customers. Some writers confine customer costs to commercial expenses plus carrying charges on meter and service invest-

17 See p. 71, *infra.*

18 Somewhat similar conceptions of customer costs are frequently found in analyses of electricity costs. Though differing as to details, the following allocations are illustrative: L. R. Nash, *P. U. Rate Structures*, pp. 240-263; N. R. Gibson, "Variable Elements in the Cost of Distribution of Electrical Energy," *Edison Electric Institute Bulletin*, I (1933), 89-90.

19 T. C. O'Hara, testimony before the Special Master, April 17, 1925, U. S. Circuit Court of Appeals (Second District), U. S. Light and Heat Corp. *v.* The Niagara Falls Gas and Electric Light Co., *Transcript of Record* (1930), p. 260; W. G. Raymond, *The Public and its Utilities*, p. 97; M. G. Glaeser, *Outlines of Public Utility Economics* (New York, 1927), p. 653; C. L. Seavey, "The Disputed Service Charge," *P. U. Fortnightly*, VI (1930), 592.

ment. Others add to these the fixed charges on a portion of the distribution lines or mains and a large share of general expenses. The divergence of opinion parallels that which is met in the definition of customer costs as those varying according to the number of customers, and in fact some writers use the two definitions interchangeably.

In contrast to the broad view of customer costs summarized above, another group of definitions rests upon a narrow view of customer costs. Accordingly, customer costs are said to be those which are traceable to the individual customer, as such, and which arise from the fact that he chooses to be connected with the company's system.[20] The investment allocable to customer cost is usually restricted to the property devoted to the customer's own use, such as meter and service line. Expenses include mainly commercial and utilization expenses. Other costs that are included under the broad view of customer costs, are here excluded; for example, all or part of the carrying charges on distribution mains. On the other hand, certain costs are included under the narrow view that are excluded under the broad view: the entire investment in meters and services is allocated to customer cost instead of that which is the equivalent of meters and services of minimum capacity. The narrow view leads to a customer cost per customer varying according to the investment in meters and services of different sizes.[21]

20 H. C. Attwill, *Report of N. Y. Com. Rev. P. S. C. Law*, II, 774; *Re* Springfield Gas Light Co. (Mass. D. P. U.), P. U. R. 1929 A 230; *Re* Malden and Melrose Gas Light Co. (Mass. D. P. U.), P. U. R. 1929 C 545, 548; Alabama P. S. Com. *v.* Alabama Power Co., P. U. R. 1925 A 413, 424; M. L. Cooke, *On the Cost of Distribution of Electricity to Domestic Consumers* (Philadelphia, 1929), p. 21; J. E. Allison, "Rate Calculations for Electric Light and Power," *Report of St. Louis Public Service Commission to the Municipal Assembly on Rates for Electric Light and Power* (1911), Appendix A, p. 6; Watkins, *Electrical Rates*, p. 69. This conforms to recent usage by the Wisconsin Public Service Commission; see H. Zinder and W. E. Caine, "Regulation of Municipal Electric Utility Rates in Wisconsin: 1931-34," *Journal of Land and Public Utility Economics*, LIX (1935), 22, and *Biennial Report of the Public Service Commission of Wisconsin*, 1934, p. 13.

21 See, Chap. VII, *infra*.

Though the extreme views of customer costs are either broad or narrow, there are many shadings of opinion and practice that fall everywhere between these two poles. To differentiate all of these would be well-nigh impossible, and, moreover, fruitless. Definitions given by writers do not always disclose clearly and precisely which costs they deem to be customer costs. Frequently, writers who define customer costs as those depending upon the number of customers, restrict these costs to those definitely and easily traceable to customers. In practice, the costs that are classified as customer costs depend upon individual opinion, the completeness of the data at hand, and the wide variety of forces entering into the making of any rate, such as law, public opinion, and business expediency.

Specific Items Entering Customer Costs

In the main, the specific items of customer costs in the gas industry are the same as those in the electric light and power industry. The chief item that sometimes occurs in the latter, but not the former, is transformer costs, *i. e.*, fixed charges upon the investment in transformers and the expenses of maintaining and operating them. It would serve no great purpose, therefore, to enumerate the items of customer costs for the two industries separately.

An attempt will be made here to review the items that have been called customer costs in one place or another. Most of these appear in the summaries of gas customer costs in Tables 6 and 7. The data are taken from several of the reports of the Rate Structure Committee of the American Gas Association,[22]

22 See Ewald Haase, "Cost of Service and Rates Based on Same," Appendix A, Report of Committee on Rate Structure, *A. G. A. Proceedings*, V (1923), 129-152, which is an analysis for the Milwaukee Gas Light Co. for 1922; Report of Rate Structure Committee, "Principles of Rate Making for Gas Companies," *A. G. A. Proceedings*, VIII (1926), 115-116, which is an analysis for a company serving 76,000 customers, all of one class; Report of Rate Structure Committee, "Report of Subcommittee on Practical Rate Making," *A. G. A. Proceedings*, X (1928), 92-101, which is an analysis for a gas company having 39,000 residential customers and 1,000 commercial customers.

and from analyses presented on behalf of the Brooklyn Borough Gas Company and the Brooklyn Union Gas Company in rate proceedings before the Public Service Commission of New York.[23] These analyses may be used as a point of departure for a study of the anatomy of customer costs. They can be regarded as illustrative of a broad view of customer costs as expounded by the American Gas Association.[24]

The percentages of various accounts that were apportioned to customer costs in the five studies are summarized in Table 6. Other cost groups—commodity or output, and demand—are

23 Exhibits 5 and 7a, submitted by the Brooklyn Borough Gas Co. in Case No. 4458 (N. Y. P. S. C.), reported as Customers for Gas in the Thirty-first Ward, Borough of Brooklyn *v.* Brooklyn Borough Gas Company, P. U. R. 1929 D 433, and P. U. R. 1931 D 129; Exhibits 54, 31 and 36 submitted by the Brooklyn Union Gas Co. in Case No. 4832 (N. Y. P. S. C.), reported as *Re* Brooklyn Union Gas Co., P. U. R. 1929 D 171 and P. U. R. 1931 D 129.

24 The cost studies of the A. G. A. Rate Structure Committees have had a wide influence in the gas industry. Some gas companies have merely adapted the A. G. A. method to their own businesses, but others have gone so far as to apply to their own data the percentages used by one or another of the Committees. Thus, the Brooklyn Borough Gas Co. allocations are based upon the 1926 A. G. A. cost study, with modifications in accordance with the experience of the company's executives; (see testimony of Mary E. Dillon, Customers *v.* Brooklyn Borough Gas Co., Case No. 4458 [N. Y. P. S. C.], *Stenographic Minutes*, pp. 489-490). A three-part cost analysis by the Boston Consolidated Gas Co. in support of a service charge was made according to the standard practice of the A. G. A. "since it represents the combined opinion of the best informed men of the gas industry." (Testimony of Dana D. Barnum, Public Hearing before the Massachusetts Department of Public Utilities, D. P. U. 3031 [December 7, 1927], p. 8). As a result of these and similar adaptations of the reports of the A. G. A. Rate Structure Committees, the A. G. A. analyses have often been referred to as the "A. G. A. formula" for the service charge. It is maintained by representatives of the gas industry, however, that it was not intended that the allocations made by the A. G. A. Committees were to be used by gas companies without modification. Instead, it was urged that each cost study by a gas company should represent the exercise of executive judgment in the light of all the relevant facts, and should stand upon its own footing and merits. ("Report of the Rate Structure Committee," *A. G. A. Proceedings*, X [1928], 93; *ibid.*, XV [1933], 36; C. S. Reed, "Who the Small User Really Is," *P. U. Fortnightly*, X [1932], 80.)

not shown here. The weight of each item in the total customer cost is shown in Table 7.

Capital Investment Items

Though the method of valuing property undoubtedly affects the amount of customer costs, it is assumed here that the value

TABLE 6

Summary of the Percentages of Capital and Operating Costs Apportioned to Customer Cost in the Gas Industry

	A.G.A. 1928	A.G.A. 1926	A.G.A. 1923	Brooklyn Borough 1926	Brooklyn Union 1928
Capital items					
Land	9.8%	0.7%	12.7%	9.6%	8.5%
Plant & equipment	15.9	0.4	2.0	15.0	3.9
Transmission and distribution	92.5	63.3	68.0	89.6	70.7
General equipment	55.0	44.6	54.5	61.3	37.7
Overhead and undistributed costs	64.0	—	38.7	61.7	37.7
Materials and supplies	—	—	12.8	—	—
Working capital	20.0	37.8	—	46.1	29.3
Going value	64.0	—	—	60.0	—
Total capital	60.6%	38.8%	35.2%	60.3%	37.1%
Operating expenses					
Production	[a]	—	—	—	—
Transmission and distribution	88.9%	61.1%	71.3%	88.0%	70.9%
Commercial	100.0	100.0	100.0	100.0	100.0
New business	25.0	30.5	50.9	40.0	—
General and misc.[b]	38.9	42.5	41.5	40.8	34.8
Taxes	30.4	36.6	22.0	53.1	36.4
Uncollectible bills	20.0	40.0	100.0	40.0	31.1
Cr.—Misc. oper. rev.	—	—	—	*100.0*	—
Total operating expenses	26.9%	27.1%	19.7%	28.4%	29.4%
Total expenses and return[c]	35.9%	30.5%	23.6%	36.9%	31.1%

[a] Less than 0.1%.

[b] Includes Retirement Expenses.

[c] Return calculated at a rate of 8%, except for the Brooklyn Union Gas Company, which applied a rate of 5%, representing net income earned on property value.

TABLE 7

PERCENTAGE ANALYSIS OF THE COMPONENTS OF GAS CUSTOMER COSTS

	A.G.A. 1928	A.G.A. 1926	A.G.A. 1923	Brooklyn Borough 1926	Brooklyn Union 1928
Return on capital					
Land	0.2%	(a)	1.4%	0.2%	0.2%
Plant and equipment	2.9	0.1	0.7	2.7	1.2
Transmission and distribution system	32.1	33.9	30.4	30.5	23.4
General equipment	0.7	0.5	0.8	0.7	0.3
Overhead & undistrib.	1.4	—	2.5	1.4	0.1
Materials & supplies	—	—	0.8	—	—
Working capital	1.1	2.5	—	2.5	1.7
Going value	5.5	—	—	5.1	—
Total return on capital	43.9%	37.0%	36.6%	43.1%	26.9%(c)
Operating expenses					
Production	(a)	—	—	—	—
Transmission and distribution system	14.1%	14.8%	23.9%	13.6%	26.5%
Commercial	21.2	18.2	14.4	20.7	25.9
New business	1.3	1.9	1.9	2.0	(b)
General & miscellaneous	12.8	18.8	11.7	13.2	11.1
Taxes	6.4	8.7	11.3	10.9	9.2
Uncollectible bills	0.3	0.6	0.2	0.5	0.4
Cr.—Misc. oper. rev.	—	—	—	*4.0*	—
Total operating expenses	56.1%	63.0%	63.4%	56.9%	73.1%
Total customer cost	100.0%	100.0%	100.0%	100.0%	100.0%

(a) Less than 0.1%.

(b) Allocated to commodity cost.

(c) Return in this instance based on net income actually earned, approximately 5%. If return were placed at 8%, operating expenses would be only 63.1% and return 36.9% of total customer cost.

of the property is not in question. InTables 6 and 7, only the return on the property is placed under capital costs. The items of taxes, depreciation, and insurance are treated here as operating expenses. In some analyses, however, they are lumped together with the return, to form "carrying charges" on investment.

The first two items listed in Table 6, namely, land, and plant and equipment, represent that part of general office land and buildings which is used for keeping customer records, making and collecting bills, selling and advertising the service, and executive control relating to customers. The part of land and plant which goes to customer cost is ascertained by estimate and observation. This seems proper, although sales and display space should be allocated to the cost of merchandising. In practice, it is difficult to apportion the land and plant in any way that is not arbitrary, and some writers therefore exclude this entire item from customer costs.

Transmission and distribution system. Of all capital accounts, this group bulks the largest in the five gas-cost analyses. It is significant that, with one exception, the item of mains was more than 50 per cent of the total transmission and distribution investment allotted to customer costs. This is shown in Table 8.

TABLE 8

PERCENTAGE ANALYSIS OF TRANSMISSION AND DISTRIBUTION SYSTEM CAPITAL ALLOCATED TO CUSTOMER COST

	A.G.A. 1928	A.G.A. 1926	A.G.A. 1923	Brooklyn Borough 1926	Brooklyn Union 1928
Mains	56.3%	46.2%	58.3%	56.6%	53.4%
Services	22.9	29.3	18.7	23.7	27.9
Meters and installation	20.6	22.7	23.0	19.4	17.4
Gas appliances	0.2	1.8	—	1.3	1.3
Total allocated to customer cost	100.0%	100.0%	100.0%	100.0%	100.0%

In the 1928 A. G. A. analysis, 90 per cent of the investment in mains was a customer cost item, and only 10 per cent a distribution demand item. In the 1926 A. G. A. study, 42.8 per cent of main investment was charged to customer cost; in the 1923 study, the allocation was 58.7 per cent; in the Brooklyn Borough case it was 88 per cent; in the Brooklyn Union analysis it was 58.1 per cent. The reason given for the allocation of such a large part of main investment to customer cost is that generally the length of mains varies directly with the number of customers served. If it were not for variations in the distribution demand, the distribution system could be made up wholly of mains of the smallest size customarily laid in a given territory. Hence, the investment corresponding to that in mains of minimum size is called a customer item, and the remaining investment in mains is considered a demand item.[25] This way of apportioning mains loads them heavily upon the customer cost group. If mains were omitted altogether from customer costs in the five studies at hand, these costs would be cut down by an amount ranging from 12.5 per cent to 18 per cent of the total.

In the electrical industry, a part of investment in distribution lines and transformers is often allocated to customer costs for the same reason. Usually, that part of the distribution lines which is devoted solely to industrial or special patrons is not counted a customer cost, nor is that part of the underground wires and conduits which is required mainly because of the physical conditions of individual or community service. Thus N. R. Gibson, in one study, charges 20 per cent of the investment in overhead lines and poles (exclusive of industrial lines) to customer costs,[26] and Nash, in an illustrative study, commits 50.8 per cent of overhead distribution investment to customer costs.[27] Investment in distribution substations is generally allocated to demand, or to special customers.

Customer cost sometimes includes a part of transformer investment, for the reason that when there are many small cus-

25 The variation in the five studies in the percentage of main investment allocated to customer cost is due to a variation in the proportionate investment in transfer, transmission, and distribution mains.

26 "Variable Elements in the Cost of Distribution of Electrical Energy," *Edison Electric Institute Bulletin*, I (1933), 90.

27 Nash, *op. cit.*, p. 248. This agrees closely with W. J. Greene, C. F. Schoonmaker, and C. B. Gorton, who allocate 50 per cent of poles, fixtures, and overhead conductors to customer costs. "Allocation of Electric Service Cost," *Electrical World*, LXXX (1922), 929.

tomers, there must be numerous small transformers, the cost of which is high per kilowatt of capacity. In contrast, a small number of large customers would require a few large transformers at a low cost per kilowatt of capacity. Thus transformer cost may be expressed as a fixed sum plus a certain amount per kilowatt of capacity.[28] The former would be a customer, and the latter a demand, element.[29] If, however, transformer investment is largely determined by collective demands and customer density, there is no reason to allocate any important part of it to customer costs. Except for rural rates, public utility commissions rarely count transformer cost as a customer cost.[30]

Services. Next to mains, the investment in services usually bulks largest in the capital items allocated to the gas customer costs as shown in Table 8. In most cost analyses, the entire investment in services is treated as a customer cost item, on the ground that it is devoted entirely to the individual customer's use. If, however, the reasoning of the A. G. A. Rate Structure Committees on the allocation of mains is followed consistently, the cost of the service pipe in excess of the cost of service pipe of minimum size should go to distribution demand. But even

28 Nash, *op. cit.*, pp. 242, 243, allocates approximately 47 per cent of total transformer investment to customers; Gibson, *loc. cit.*, assigns 20 per cent, and Greene, Schoonmaker and Gorton, *loc. cit.*, assign 10 per cent.

29 The amount of transformer investment allocated to customer cost varies considerably. In a rural area, where each customer has a separate transformer, there is good reason to allocate the entire transformer to customer cost. In urban areas, the density of population reduces the amount of transformer capacity required per customer, but even here the inclusion of transformer investment would increase the customer cost considerably. R. T. Livingston, in a breakdown of distribution costs on a per customer basis (without reference to customer or demand components), places average transformer investment per domestic customer at $15.50, in contrast to $8.00 for services and $12.50 for meters. (R. T. Livingston, "How Cost of Distribution Figures in Electric Bills," *P. U. Fortnightly*, XIII (1934), 167.) Nash's allocation, *op. cit.*, p. 288, produces a transformer investment of $7.29 per customer as customer cost.

30 One exception is Maires *v.* Flatbush Gas Co. (N. Y. P. S. C., 1st Dist.), P. U. R. 1920 E 930.

if this demand component in services is singled out, it is relatively small; in the 1923 A. G. A. analysis it amounted to 13.1 per cent of the entire investment in services. In the electrical industry, as a rule, services are apportioned wholly to customers. This has been done even though the minimum capacity principle has been applied, in the same study, to distribution lines, transformers, or meters.[31] And, as in the gas industry, even though the demand component in services is singled out it is small.[32]

Meters. The investment in a meter, plus settings and connections, is considered to be devoted solely to the use of the individual customer. Most domestic gas customers are adequately served by a five-light meter, but those with a larger demand need a larger meter. Part of the investment in meters, therefore, is due to demand. The demand element may be set apart by assigning the cost of meters in excess of the cost of minimum (five-light) meters to distribution demand.[33] This was done in four of the cost analyses recapitulated in Table 6. Yet the fact that most customers are served by meters of minimum size, and that the investment per unit of meter capacity decreases as the size of the meter increases, served to keep the percentage of the meter investment allocated to distribution demand at a low figure—8 per cent in the 1928 A. G. A. study, 11.7 per cent in the 1923 A. G. A. analysis, 16.6 per cent for the Brooklyn Borough Gas Company, and 20 per cent for the Brooklyn Union Gas Company. For electricity supply, the demand component of total meter investment seems to run in the same magnitude, ranging from 5 per cent to 20 per cent.[34] Where all customers are served by meters of identical

31 Gibson, *op. cit.*, p. 89; Nash, *op. cit.*, p. 242.

32 Greene, Schoonmaker, and Gorton, *loc. cit.*, p. 929, allocate 10 per cent of services to demand and the remainder to customers.

33 A. G. A. Rate Structure Committee, *A. G. A. Proceedings*, X (1928), p. 96.

34 Nash, *op. cit.*, p. 242; Greene, Schoonmaker, and Gorton, *loc. cit.*, p. 929.

size, there is no doubt that meter investment is a customer item in its entirety.

Appliances. The small investment in installations on the premises of customers is allocated wholly to customer costs. This is proper when appliances installed on customer premises are owned by the company, but where installations are not uniform or are infrequent, some slight discrimination may result from averaging them on a per customer basis.

Street lighting equipment. The report of the A. G. A. Rate Structure Committee for 1926 states that all of street lighting investments should be allocated to the customer cost group.[35] There is some ground for this view if only one class of customers is served and street lighting investment is small. But if several classes of customers of different consumption characteristics are served by a company, and if an extensive street-lighting system serves commercial as well as residential territory, the costs of street lighting should be separately ascertained and billed.[36]

General equipment. This group of accounts includes equipment which serves a general purpose, such as that used in general offices, stores, shops, transportation, and laboratories. In the five gas-cost analyses now under scrutiny, each type of equipment is allocated to customers, output, or demand in accordance with the use to which it is devoted. Although from 37 per cent to 61.3 per cent of all general equipment is allocated to customer cost (as shown in Table 6), the return on the investment so allocated is less than 1 per cent of total customer costs. In most gas and electric cost calculations, the investment in general equipment is lumped together with office investment under the heading of general capital. The mode of allocation varies, but ordinarily as much of general capital as can be traced to each cost group on the basis of use is assigned, and the remainder is distributed in proportion to the capital or expense allocations already made.

35 *A. G. A. Proceedings,* VIII (1926), p. 104.

36 Haase, *A. G. A. Proceedings,* V (1923), p. 138.

Intangible capital items. Customer cost often includes a part of the return on intangible capital items, such as legal expenses, taxes and interest during construction, organization expenses, as well as franchises and going value. These are sometimes allocated to customer cost in proportion to the apportionment of the tangible fixed capital.[37] Thus, the 1928 A. G. A. analysis puts 64 per cent of going value in customer costs, and the Brooklyn Borough study allots 60 per cent of going value to the same group. The return on going value to be collected as a customer cost was not to be despised in either case; as is shown in Table 7 it constituted 5.5 per cent of customer cost in the former case, and 5.1 per cent in the latter.

It is hard to see any real connection between intangible capital and customer costs except for those elements which represent actual expenditures in attaching customers to the system, and retaining them. Going value does not increase if the company is so inefficiently managed that customer costs are inordinately high. In the majority of cost analyses presented in rate cases, intangibles are not mentioned as customer costs. But their inclusion is, of course, consistent with the A. G. A. theory of customer costs.

Working capital. Working capital is allocated on the basis of the uses made of it. Table 6 shows that from 20 per cent to 37.8 per cent of return on working capital was allocated to customer cost, but this comprised less than 2.5 per cent of total customer costs in all cases. Working capital also may be apportioned on the basis of the distribution of fixed capital. Any method of allocating it may be applied so as to be arbitrary; yet it seems just that some share should be loaded upon customer costs.

Summary of capital cost allocation. Customer costs are strongly influenced by the method of apportioning capital costs.

37 Occasionally, some items, such as franchises, are allocated entirely to customer costs. For example, see, C. H. Dauchy, " Costs in Rate-Making," *Electrical World*, XCIII (1929), 733; Greene, Schoonmaker, and Gorton, *loc. cit.*, p. 929.

In the five studies summarized in Table 7, the return on customer capital was about one-third of total customer costs. Furthermore, it must be noticed that the allocation of certain operating expenses is guided by the allocation of capital items. Such expenses are retirement expense, taxes, insurance, maintenance, and supervision. True, the minimum-capacity basis of allocating investment represents an extreme view and is not universally accepted. The more common view is that customer investment is limited to meters, services, and sometimes a share of general office space and equipment. It is fruitful to inquire further into the methods of allocating capital items.

Does the investment in distribution mains and transformers vary directly with the number of customers attached to a utility system? The A. G. A. Subcommittee on Stand-by Charge Rates argued that as the number of customers grows so does the distribution investment. This, however, is far-fetched. If a company opens a new territory and adds 10,000 customers to its rolls, it must make a substantial investment in distribution mains; in contrast, if 10,000 customers are added in territory already laid out and developed, no new mains may be needed. The layout of the distribution system is profoundly affected by the density of customers, and the character and quantity of their demand for gas or electricity. If the entire investment in the distribution system is allocated to customer capital, and is spread equally among customers, the ensuing cost per customer and the rate based upon it will probably discriminate against the customer with a small demand and a good load factor, in favor of the customer with a high demand and a poor load factor. The argument that mains are a customer cost goes too far. It could as well be said that the entire investment in all facilities is a customer cost since the sole reason for investing is to supply customers!

Those who allocate to customer costs only the costs related to the distribution system of minimum capacity, have a much better case than the one above. They argue that a large part of the distribution investment is fixed by the extent of the system, which in turn largely depends on the number of consumers. Hence it seems just to make each customer bear directly a part of the costs of the distribution system; for example, costs on that part of mains equivalent to the average number of feet of main which the consumer uses, based upon the smallest pipe laid.[38]

But the contrary view, that no portion of distribution lines or mains (and, in the case of electricity, transformers) should be classified as customer costs, seems to be equally reasonable. If a company puts in a main of the smallest size, that main may be laid for a single large customer or for a dozen smaller customers. There is reason to believe that the cost of street mains is a function of the collective demand of all customers, and that it is regulated by the area of the territory and the density of customer population and demand, as well as by the number of customers. Mileage and diameter of mains, which govern distribution system costs, depend upon the area served and the demand. If demand in a new territory were not large enough, a gas company would not carry even a two-inch main into that area, no matter what the number of customers. If, however, one large customer came with a huge demand, the gas main would go in promptly. Consequently it has been argued that the fixed charges upon mains, and their maintenance, should be allocated to demand costs rather than to customer costs.[39] Some authorities would go so far as to exclude service pipes and wires. For instance, Watkins says:

38 Milwaukee *v.* Milwaukee Gas Light Co., 12 Wis. R. C. R. 441 at 478, 479 (1912); W. G. Vincent, *Pacific Coast Gas Assoc. Proceedings*, XI (1916), 269; "Report of Rate Structure Committee," *A. G. A. Proceedings*, III (1921), 43; Alan Cunningham, testimony before Massachusetts Department of Public Utilities, *Hearings in D. P. U. 3530* (1929), pp. 51, 52, 56.

39 T. V. Purcell and J. M. Spitzglass, Discussion of "Report of Rate Structure Committee," *A. G. A. Proceedings*, III (1921), 53, 54, 63; John Bauer, "The Effect of the Service Charge on the 'Small User'," *P. U. Fortnightly*, IX (1932), 271; Moritz *v.* Edison Electric Illuminating Co. (N. Y. P. S. C., 1st District), P. U. R. 1917 A 364.

The cost of connecting an apartment house containing fifty families with a street main is no more separable, and chargeable to the individual consumer, than are the wires stretched along a block-front to supply 20 private houses. In a city of multiple dwellings, the distribution system extends almost to the meter and its cost is a joint cost.[40]

Nor for the same reason should any element of fixed charges on the distribution system be included in the customer charge.

Such disagreement as to the allocation of capital costs, strongly suggests that the theory of cost analysis that is actually applied often is a rationalization of a desired result.

Operating expenses in customer costs. Usually, the major portion of customer costs consists of operating expenses. Production expenses do not enter customer cost at all, except for a small portion of maintenance of laboratory equipment, which might easily be included in production demand.

Transmission and distribution expense. As shown in Tables 6 and 7, these expenses add heavily to the total of customer costs. The items of distribution superintendence, supplies, and expense, which came from the operation of the street department and the fitting and repair shops, are sometimes allocated according to the division of the investment in the distribution system,[41] and sometimes according to a detailed breakdown of each account.[42] The same methods are applied to some of the maintenance accounts—such as mains, shop buildings, distribution implements and accessories, and street lamps—with substantially similar results. Ordinarily, the remaining expenses under this heading—work on consumers' premises, removing and resetting meters, maintenance of service lines or pipes, appliances, consumers' meters and installations—are placed in

40 G. P. Watkins, *Electrical Rates* (1921), pp. 89-90. A similar view was voiced by Commissioner Maltbie in a dissenting opinion in, *In re* Stadtlander *v.* New York Edison Co. (N. Y. P. S. C., 1st District), P. U. R. 1915 B 685.

41 A. G. A. Rate Structure Committee, 1926, *op. cit.*, p. 112; Haase, *op. cit.*, pp. 136, 137.

42 A. G. A. Rate Structure Committee, 1928, *op. cit.*, p. 99.

customer costs in full. But if the distribution investment to which these latter expenses relate has been apportioned on the basis of the minimum capacity required, they also contain a demand component. This may be fixed either by applying to these expenses the percentages used for the division of investment, or by an analysis of each particular account.[43]

It is sometimes suggested that each customer should pay a charge to cover utilization expenses only when these are incurred on his behalf. These costs relate to the work done on consumers' premises, and to connecting and disconnecting meters. Some customers rarely or never call for the company's repair men, or rarely cause the disconnection of service. Yet if utilization expenses are buried in a customer charge, these careful customers pay costs created by neighbors who do not pay their bills, or who demand frequent repairs on faulty equipment. It is easy to determine a charge for connections or disconnections, but such a charge is often difficult to collect. A separate charge for servicing appliances might cause customers to take better care of their equipment; this would reduce utilization expenses. On the other hand, it is not easy to find the cost of each job of servicing, and a charge for such work would be hard to collect. It would, moreover, keep customers from asking to have defective equipment examined and serviced. The lack of prompt and expert servicing would impair the quality of the service, and would lead customers to hesitate to install new appliances. It has been easier to recover utilization expenses as a part of the customer charge than to collect a separate fee for repairs or adjustments. It seems best, however, to include consumer job costs in commodity or energy cost. Otherwise, with utilization costs buried in the customer charge, a customer with little equipment and a small consumption pays just as much utilization costs as does one with much equipment and a large consumption.

43 A. G. A. Rate Structure Committee, 1928, *op. cit.*, p. 99; Haase, *op. cit.*, p. 136; Nash, *op. cit.*, p. 243.

Commercial expenses. As a rule, commercial expenses are charged entirely to customer costs.[44] They constitute from 15 per cent to 25 per cent of total customer costs in the gas-cost studies listed in Table 7. In many rate cases, commercial expenses are the back-log of the customer costs presented by a company in support of a service charge.

New Business Expense. This consists of the expenses of advertising, demonstrations, canvassing and soliciting, and related outlays. There is no general rule for the apportionment of new business expense. Some analysts rely upon familiarity with operating conditions in order to estimate the proper share of new business expense to be borne by customer costs.[45] Others apportion new business expense on the basis of the apportionment of all other costs, including return upon property,[46] or use some variation of this method.[47] As a result of different methods, the proportion of new business expense going to customer costs varies from all to none.[48] In the five gas-cost studies

44 Some accounts, such as Commercial Contracts, do not pertain to ordinary customers, and should be charged to a specific class of customers; so also should a related portion of Commercial Supplies and Expense. See, A. S. B. Little, "Basis of Expense Allocation for Customer and Commodity Cost," Exhibit 28, Customers *v.* Brooklyn Borough Gas Co., Case 4458 (N. Y. P. S. C., 1928); N. R. Gibson, *E. E. I. Bulletin*, I (1933), p. 90.

45 Haase, *op. cit.*, p. 137.

46 A. G. A. Rate Structure Committee, 1926, *op. cit.*, p. 113.

47 Nash, *P. U. Rate Structures*, pp. 243, 251.

48 See, Little, *op. cit.*, p. 4; *Re* Grand Rapids (Mich. P. U. C.), P. U. R. 1923 C 453, 520, 521. The old Public Service Commission of the First District of New York disallowed new business expenses and uncollectible bills as a customer cost in Moritz *v.* Edison Electric Illuminating Co. (P. U. R. 1917 A 364). The Commission argued that as the proposed service charge was intended to prevent the serving of small customers at a loss, it was illogical to place the burden of such expenses on unprofitable customers which the company presumably should not solicit. The Commission, however, overlooked the requirement of public policy that all customers in populated districts be given service, and the possibility that such customers might develop into large consumers. There is a much stronger reason for disallowing new business expenses, at least in part, when they are not effective owing to the high price of gas, or to depression conditions. (See, *Re* West Ohio Gas Co. [Ohio P. U. C.], P. U. R. 1933 B 433.)

presented in Table 6, the portion of new business expense charged to customer costs ranged from nothing to one-half.

If customer cost is looked at narrowly, only that portion of new business expense which is devoted to getting ordinary customers to take service should be charged to the customer cost group. The remaining new business expenses, devoted in the main to increasing the demand and consumption of customers, should be placed in demand and output costs. Yet it is hard to break down expenses as general as those for advertising, demonstration, and soliciting. Even so, the expenses of getting special customers or special business should not be included in customer cost. Similarly, the costs of handling, displaying, and marketing utility appliances should not be borne by a customer charge under the guise of new business or other expense. Merchandising and jobbing expenses ought not to enter rates at all. These expenses should be charged to a special account, distinct from other promotional expenses, as has been required by the Public Service Commission of New York[49] and the District of Columbia Public Utilities Commission.[50]

General and miscellaneous expenses. Certain items of general expense, such as retirement, insurance, stores, and transportation expenses, are properly distributed upon the basis of the physical property to which they pertain. The remaining items, which are treated in a more or less arbitrary fashion, consist of administrative and other general office salaries, miscellaneous general expenses, maintenance of general structures, legal expenses, undistributed adjustments, rentals, injuries and damages, regulatory commission expenses, and expenditures for relief and welfare work. Because of their mixed nature, and because they apply to all operations of a utility, these expenses cannot be traced directly to customer, demand, or output costs. Consequently, some more or less arbitrary method of division must be used. One way is to estimate the share charge-

49 *Re* Uniform System of Accounts for Electric and Gas Companies, P. U. R. 1933 B 448.

50 *Re* Accounting for Gas Companies, P. U. R. 1931 B 436.

able to each cost group on the basis of an examination of the items entering these several expense accounts.[51] Another way of treating undistributed general expenses is to assign them in proportion to the amount of expenses already distributed to customer, demand, and output costs. And, again, a combination of the two foregoing methods may be used.[52] Infrequently, general expenses are placed entirely in the customer group.[53] This procedure may be equitable in some instances, but it is hard to see that all of general expense should be loaded upon customer costs as a rule.[54] Some experts say that any apportionment of general and miscellaneous expenses is so speculative that no part of it should go to customer cost.[55] Nevertheless, it can hardly be denied that some general expenses are directly related to the mere fact of the continuation of customer connections. But any attempt to ascertain the extent of such a relationship will be more or less arbitrary.

Taxes. Property and personal taxes may be directly assigned to cost groups on the basis of the allocation of capital to which they refer. Sales taxes and taxes on income, are more difficult to apportion. In some cases income taxes are distributed on the basis of the apportionment of all other items, including return on property.[56] In other cases they are placed

51 See the analyses of the A. G. A. Rate Structure Committees, 1923, 1926, 1928, *op. cit.* The data on gas costs presented in Tables 6 and 7, show that about 35 to 40 per cent of general and miscellaneous costs are traceable to customer costs, and that these constitute from 11.1 per cent to 18.8 per cent of total customer costs.

52 Nash, *op. cit.*, pp. 240, 243, 260. 53 Gibson, *op. cit.*, p. 90.

54 In an old study, Allison showed that the relationship between general expenses and output was much more direct than that between general expenses and number of customers. But the results were thought to be inconclusive because of the variation in the accounting methods of assigning specific items to general expenses. See, James E. Allison, " Rate Calculations for Electric Light and Power," *Report of the St. Louis Public Service Commission to the Municipal Assembly of St. Louis on Rates for Electric Light and Power* (St. Louis, 1911), Appendix C, pp. 13, 14.

55 Little, *loc. cit.*

56 A. G. A. Rate Structure Committee, 1928, *op. cit.*, p. 100.

entirely in output costs. If, however, income taxes and sales taxes are considered to be a function of revenues, it is clear that they should not add to customer costs, except to the extent that a customer charge yields revenues. Frequently, all taxes are expressed as a percentage of property investment and in this way they enter customer costs. But this is accurate only so far as the sum of taxes is made up of property taxes. The importance of taxes as an item of customer cost is shown in the data on gas customer costs (Tables 6 and 7). From 20 per cent to 30 per cent of all taxes paid (not including sales taxes) went into customer costs, and constituted from 6.4 per cent to 11.3 per cent of such costs.

Uncollectible Bills. This item is often classed as a general expense and is treated as such. In some cases it has been treated wholly as a customer cost, and in others, wholly as a commodity cost. Probably some share of uncollectible bills varies with customers, but any allocation is bound to be quite arbitrary. The matter is not of great consequence. In the five gas-cost studies shown in Table 7, uncollectible bills made up from 0.2 per cent to 0.6 per cent of customer costs.

Gas or energy losses. An item which sometimes is placed in customer costs is that of losses of energy or gas in the distribution system. In the case of electricity the amount of such losses includes transformer-core losses and meter potential coil losses.[57] Thus sometimes energy losses in meters are held to be customer costs.[58] And if a part of transformer investment and operation has been assigned to customer cost, a part of transformer core losses also should be so assigned. Customarily, however, the cost of energy losses is absorbed in the

57 Nash estimates that in a well-designed residential distribution system, loss of energy in meters amounts to 4.0 per cent of kw.-hr. sold, in secondaries and services 1.0 per cent, and in distribution transformers 5.0 per cent. System energy losses, of course, would take in those in the remainder of the distribution and transmission system, and the percentages would decrease as load factor increased. Nash, *op. cit.*, p. 246.

58 Watkins, *op. cit.*, p. 69; Greene, Schoonmaker, and Gorton, *op. cit.*, p. 939.

energy output cost, and an attempt is made to make each class of service bear its share of the cost of energy lost, as far as that can be estimated.

A good part of operating expenses may be placed in the customer cost group solely because a part of capital has been so allocated. If the return on the investment in electric transformers and lines (or in gas mains) is disallowed as a proper customer cost, the operating expenses that were allocated to customer cost on the basis of the capital cost apportionment should also be disallowed. Portions of distribution superintendence, distribution supplies and expense, maintenance of shops and buildings, distribution implements, retirement expense, insurance and taxes—all of which followed the ratios applied to distribution capital—should be dropped from customer cost. In addition, certain other items which found their way into customer cost by the method of allocating general accounts in proportion to the expenses already distributed should be stricken out of customer costs; these items would include part of new business expense, uncollectible bills, and general and miscellaneous expenses.

Conclusions

The broad view loads a large part of distribution investment and expense on customer costs. The amount per customer determined in this way is much larger than that fixed according to a narrow view. For example, fragmentary data show that if the broad view dominates a rate analysis, customer costs in the gas industry are about \$1.00 to \$1.50 per month.[59] In contrast, if customer costs are narrowly confined, the costs per customer are about 50 or 60 cents per meter per month. In the electrical industry the contrast is not so great, but a wide spread is found nevertheless.

A large service charge which follows from the broad view makes it possible for a company to have lower follow-on rates than would be possible with a small service charge. This does

59 See Appendix A for a study of the amount of customer costs.

not mean that the company will actually reduce rates to the greatest possible extent; it will not do so if the amount of the reduction is likely to have a limited promotional effect. Moreover, whether or not a broad view of customer costs may be applied without discrimination among customers depends upon the nature of the territory served. It may be possible to do so where customers are uniformly distributed over the territory, and when the character of their demands is also uniform. But many companies serve territories in which the density of customer population, and the time and amount of demand vary widely. Such conditions make it difficult to devise a customer charge based on a broad customer cost which does not cause excessively complicated rate schedules.[60]

The main defect of the broad view is that it may easily be pushed so far as to lead to arbitrary and discriminatory rates. If the broad view is valid, why should not all distribution investment and expense be divided equally among customers? This is practically what is done when some rate experts take the average distribution cost per customer and divide it by the number of kw.-hr. consumed by various customers. In this way, it is easy to show that the average price actually paid per kw.-hr. by the small customer is far less than the average cost per kw-hr. as so determined.[61] But this assumes that there is no material variation in customer density, and that the demand of all customers is alike. Obviously a customer with a large demand will be charged less than his share of demand cost. If he has a large consumption, the average distribution cost is spread over a large kw.-hr. base; this results in a low average cost per kw.-hr. which in turn is held to justify the low rates necessary to get his patronage in competition with alternative services. It may be said, of course, that no rate analyst ever pushes the broad view to the extent of including

60 See pp. 169 ff., *infra* for a discussion of service charges based on customer density.

61 See, for instance, E. Gruhl, "Discrimination," *The Annals* of the American Academy of Political and Social Science, CLIX (1932), 33-34.

all distribution costs. The reason for this, however, is not because the necessary rationalization is lacking, but because it would be difficult to collect a very large service charge. This might not be true in an exceptional case, as where all customers take such large amounts of service that the service charge itself would be small in comparison to the entire bill.

The very fact that the customer costs can be placed at any commercially feasible level if the broad view is used, seems to justify an "empirical" approach to the service charge. It can be argued that since a main objective of rate-making is a promotional rate schedule, the amount of the service charge should be fixed with that in mind. The best procedure would be to assume several hypothetical service charges of various amounts and then to adjust the follow-on rates for each of these so as to yield the total revenues required for the business in question. The best combination would be that which would best exploit the market. The higher the service charge collected, the lower the follow-on rate, and the more promotional the schedule. Of course the amount of the service charge would have to seem reasonable to the consuming public.

Several weaknesses make such an "empirical" method unacceptable if used alone. If the service charge covers more than the minimum customer costs of any type of consumer, some customers pay more than their proper share of customer costs. Also, the customer costs paid by each customer are not what the customer, as such, actually costs the company, but what he can afford and will pay for the privilege of being connected to the company's lines. True, the value of service does help to give some protection to the customer, especially where consumer opposition is politically effective and good substitutes limit the price for service. Where these conditions do not hold, or hold in small degree, the empirical approach leads to extortionate and discriminatory service charges. This illustrates the possible consequences of the broad approach to customer costs.

The main criticism of the narrow view of customer costs is that it leads to a small service charge, which, because it is not feasible to collect another fixed charge to cover demand costs, requires follow-on rates so high that the rate schedule as a whole does not promote consumption. There is force to this objection, though often demand costs may be recovered readily and equitably in the primary blocks of the commodity rate. But the objection rests on the assumption that a promotional rate which benefits a few customers is justified even though it discriminates against small customers. A rate of this kind may be equitable and desirable from the point of view of the company which seeks maximum revenues and income. It may be far from reasonable from the point of view of the customers who are forced to pay a higher service charge.

Customer Costs as Marginal Costs

The narrow view of customer costs rests on the idea that a customer should pay only those costs which he causes, though he may exert no demand or take no electricity. These costs approximate the additional costs of taking on a customer and maintaining the connection, without actually supplying him with electricity.[62] They would not be incurred if the customer were not connected to the lines of the company. The company should cover at least these costs, and might properly collect even more if in so doing it would not restrict the maximum use of service by customers. But if it did impose a service charge larger than the marginal customer cost, some customers would not ask for connections even though they would be willing to pay the additional cost. If the company avoided taking on a customer when he would be willing to pay a customer charge somewhat more than the marginal cost, but somewhat less than the average customer cost, it would be guilty of the economic crime of causing a loss of utility equal to the difference between the marginal cost and the price offer of the additional customer.

62 It is assumed here that the energy costs and the demand costs are covered by a separate charge.

Marginal cost in the short run, however, may be different from that in the long run. The addition of one customer does not require, for instance, more clerks, meter-readers, or maintenance men. But the addition of a large group of customers would call for a larger operating staff and more office space. The marginal cost then might be less than, equal to, or more than the average customer cost, depending upon the stage of development. Care must be taken not to take on patronage at a rate which does not fully cover the costs when the business is developed, unless that rate is expressly stated for the interim period. Though the immediate marginal cost of taking on one customer would probably be less than average customer cost, over a period of time it would be about the same as average customer cost. This would be especially true of the customer investment cost, for each customer would require the same meter and service investment as every other, assuming no price changes and one class of customers.

The great drawback to a service charge based on marginal customer costs, where these are not the same as average customer costs, is the difficulty of applying it. If a company were operating in a stage where marginal customer costs were less than average customer costs, revenues from a customer charge would be less than total customer costs. If the company were operating with marginal customer cost higher than average customer cost, the customer charge based on marginal cost would bring in more than enough to cover total customer costs. It would be justifiable for the government to impose an excess profits tax on the company in the latter case, and to give a subsidy in the former. Such action probably is out of the question at present. Subsidies, of course, might be derived indirectly from making energy or demand charges heavier than otherwise necessary; this would work well only when such charges would not check maximum use. These difficulties show that the average customer cost for a given group of customers must be used as a basis for the service charge. Nevertheless, the marginal analysis strongly supports the view that in cost allocations the customer cost group should be narrowly interpreted.

The contents of this chapter suggest that in formulating a service charge, weight must be given to the willingness of people to pay it. For several reasons, customers have objected to service charges. These objections will be examined in the next chapter.

CHAPTER V

OBJECTIONS TO THE SERVICE CHARGE

Consumers generally have opposed the introduction of service charges. They have done so partly because they have not understood the reasons for using it, and partly because they have not been willing to pay the larger bills which often follow a change to a service-charge rate. This resentment has been especially keen where utility companies have attempted to exact large service charges.[1] Opposition also stems from technical defects in the design of the service-charge rate, and from errors in the method by which it is introduced.

The main technical defects which may occur, and which may be readily remedied, have been discussed in preceding chapters. A service charge may be objectionable if it is heavily loaded with demand costs by a process of more or less rough averaging, which does not take account of wide differences in the demands of various customers.[2] Even though a service charge is equitably loaded with demand costs, numerous obstacles prevent its use. Almost any such service charge will be difficult to design without discriminatory features, and will seem excessive to customers.

In some instances, service charges have been opposed because they have been fixed arbitrarily. No service charge should be fixed merely to bring in a certain amount of revenue, without a clear enumeration of specific costs that are functionally related to the charge proposed. For even with the clearest and most justifiable determination of the service charge, the utility operator must face the protests of customers whose bills are increased by the service charge. The natural suspicion of such

1 M. L. Cooke attributes the widespread prejudice against the service charge to the efforts of utility companies to exact too large a charge; *On the Cost of Distributing Electricity to Domestic Consumers*, pp. 20-21.

2 This raises the question, Should the service charge be variable or constant? This problem is discussed in Chapter VII.

customers is easily fanned into open hostility if the cost basis of the service charge can be shown to be arbitrary in any important respect.

Popular Arguments against the Service Charge

Public hostility to the service charge crystallizes about certain arguments against it. It is usually said that the service charge is arbitrary in amount, that it is a charge for service not rendered, that it is not used in other businesses and hence is a monopolistic extortion, and that it is socially unjust because it increases the bills of small customers, who can least afford to pay. The first argument may be dismissed here, not as being without merit, but as depending upon the facts of the case at hand. The second also may be dismissed. True, customers get no " service " in the form of gas or electricity in return for the service charge. But they do get " service " in the sense that the company provides and maintains all the facilities that are necessary to make gas or electricity available continually, whenever and in whatever amount it is called for by the customer. Properly conceived, therefore, the service charge is really levied for service rendered; nevertheless, the voicing of this objection shows that there has not been a proper explanation of the service charge to the public. The last two objections warrant further attention, and will be treated here at length.

Is the Service Charge a Monopolistic Device?

Service charges are often opposed on the ground that ordinary businesses do not impose them, and that their use by utility companies is an abuse of monopolistic power. It is said that a storekeeper does not demand a service charge from each customer for maintaining a stock of goods and delivering merchandise. In answer, the defenders of the service charge say that the storekeeper is not compelled by law to keep a complete stock on hand ready for instant use. He is not forced to sell to any person at less than cost, nor does he have to maintain day and night delivery service for all people. The grocer, for instance, does not have to provide a scale on the premises of every customer to weigh goods upon delivery. All these costly obligations are imposed upon a public utility by law, and they are a strong reason for making each customer pay a fixed amount for the services required. The business of the ordinary merchant differs sharply from that of gas and electricity supply with respect to the character of investment and the mode of meeting the demands of customers. Thus the proponents of the service charge conclude that its absence in merchandising businesses is not proof of its inequitableness in the gas and electrical industries.[3]

A further defense of the service charge is that it is not limited to public utilities, but is in fact used in many other businesses.[4] Taxicabs have a relatively high initial charge and they charge for waiting. Banks often make a charge if a depositor's balance falls below a specific minimum. Cafes, nightclubs, and similar places of amusement often assess a " cover " charge in addition to the cost of the food and beverages served, in order to recover at least the cost of the orchestra and entertainers. Some use a minimum charge for the same purpose. Theater ticket agencies often collect a small sum for an annual membership card to cover readiness-to-serve costs. Many other examples might be cited.

Some competitive enterprises recognize customer costs in their pricing policies in a negative fashion. If a person places property insurance for a period of years, the cost per annum is usually less than for repeated transactions involving costs for additional clerical work. The same is true of subscriptions to periodicals. A season ticket to the opera costs less per performance than a single ticket. Some grocery stores refuse to

3 C. L. Seavey, "The Disputed Service Charge," *P. U. Fortnightly*, VI (1930), 595; Nash, *P. U. Rate Structures*, p. 195.

4 Most of the following examples appear throughout rate literature. See especially Nash, *P. U. Rate Structures*, p. 195; Committee on Rate Fundamentals, *Better Forms of Rates for Gas* (1927), pp. 16-17; C. S. Reed, "Are We Peculiar," *A. G. A. Monthly*, VIII (1926), 267-269, 308.

deliver packages unless the purchase exceeds a certain minimum, in order to avoid the relatively heavy costs of delivering petty purchases.

These instances show that service charges appear in other than utility industries, and that demand and customer costs affect the pricing of the service or commodity. But the critics of utility service charges say that the collection of service charges in the foregoing instances is evidence of monopoly power. It is alleged that in most of the businesses enumerated, a large element of special appeal or of monopoly power permits a price policy which could not be followed in a strictly competitive business.

Whether or not there is monopoly or special appeal is immaterial. The fact that special price policies may be followed in some non-competitive businesses does not establish a valid objection to the use of the service charge by public service monopolies. It might be argued that a service charge should be used in a competitive business as a matter of good economics.[5] But the possibility of collecting service charges in retail business is remote, owing to the competitive character of the business and to the lack of public control over, or sanction of, pricing practices. A service charge is more easily collected by enterprises which have a monopoly, or control a restricted market, than by competitive businesses. This explains why the service charge is or is not used, but it does not show that the service charge is not justifiable. Of course, it is possible for a public utility to use it to levy heavily upon those who cannot protect themselves by resort to alternative services. Extortion, however, can be checked by careful analysis of costs. Due care must be taken to avoid indiscriminate averaging of overhead costs in the service charge, which might thus become an instrument of monopolistic oppression.

5 The use of service charges in competitive business might well cause important changes. Competition might be intensified as to some prices; some stores might find it more difficult to exist; price structures might be made more rigid; efficiency of operation might be increased if patronage were

WHO IS THE SMALL CUSTOMER?

A common and persistent objection to the service charge is that it increases the bills of small customers who, because of their low incomes, are least able to pay an increased monthly charge for gas or electricity.[6] In answer, it is vigorously maintained that the small users whose bills are increased by a service charge rate usually are well able to pay such an increase.[7] It is contended that small or so-called "casual" or "convenience" users of gas or electric service, far from being poor

concentrated. Service charges in industrial or commercial prices would have far-reaching and possibly unpopular effects.

6 Witness the solicitude expressed in a statement as to the effect of a service charge in New York City, that "... it must be borne in mind that it is the small consumer who can least be so burdened and that, as in all matters of government, they must and should be assisted by the larger consumers to whom this slight difference would be negligible"; *Brief of Silvian Palmer of Interboro Electric Light and Power Company, representing William Randolph Hearst, United Cigar Stores of America, and other large customers of the New York Edison Company and United Electric and Power Company*, filed with the New York Public Service Commission in Case No. 6367 (1931), p. 4.

7 The main authorities voicing this point of view (which is more fully developed below) are:

W. L. Ransom, *The Service Charge as a Part of the Rate for Gas* (1921), p. 15; "Report of Committee on Rate Structures," *A. G. A. Proceedings*, V (1923), 84, 153; W. G. Raymond, *The Public and its Utilities*, p. 125; H. C. Spurr, *Guiding Principles of Public Service Regulation* (1926), III, p. 424; A. G. A. Committee on Rate Fundamentals, *Better Forms of Rates for Gas* (1927), p. 17; "Report of Committee on Public Utility Rates," *N. A. R. U. C. Proceedings*, XXXIX (1927), 448; "Report to National Directing Committee of Executives," *Report of the Rate Structure Committee of the American Gas Association* (New York: 1933), Appendix A, p. 37; W. S. Leffler, "A Study into the Prices of Domestic Electric Service," *N. A. R. U. C. Proceedings*, XLI (1929), 444-465; L. R. Nash, *P. U. Rate Structures*, pp. 154, 156; Massachusetts Department of Public Utilities, *Special Report of the Department of Public Utilities Relative to the Feasibility and Expediency of Prohibiting So-Called Service Charges in Domestic or Household Rate Schedules of Gas and Electric Companies*, Senate No. 50, D. P. U. 4693 (1934), p. 7; Alabama Public Service Commission *v.* Alabama Power Co., P. U. R. 1925 A 413, 421; *Re* Laclede Gas Light Co. (Mo. P. S. C.), P. U. R. 1929 A 591, 596; *Re* Public Service Electric and Gas Co. (N. J. B. P. U. C.), P. U. R. 1929 E 17, 26.

people, are well-to-do apartment dwellers whose mode of life requires only a casual, slight, or occasional use of gas or electric service.[8] Such people eat away from home frequently, and their home meals include many pre-cooked or prepared foods. Their washing and ironing is done in commercial laundries. Hot water is supplied by the landlord. They frequently leave home for vacations.

Other consumers of small quantities of service, especially gas, are small storekeepers, professional men, lodge halls, etc., whose consumption is difficult to increase either by lower rates or by added sales efforts; these are not overburdened by the slight increase attending the service charge. In contrast, the working-class family consistently requires larger quantities of gas and electricity for vital domestic tasks which it cannot afford to have done elsewhere or otherwise. This type of consumer uses most of the important appliances, and calls for service consistently and continuously. The defenders of the service charge conclude, therefore, that the lower-income customer is not injured by the service charge because the reduced commodity rate permits him to use more gas and electricity at little added cost. Thus the Committee on Public Utility Rates of the National Association of Railroad and Utilities Commissioners has said, " The three-part rate can well be stated as the best rate for the poor man." [9] To determine the validity of these conclusions it is necessary to examine the factual evidence on the characteristics of various types of customers.

Consumer Characteristics: The Gas Industry

Numerous statements have been made to the effect that the small consumers are not in the low-income classes, and regulatory commissions have usually accepted them as being true. Sometimes actual studies of consumer characteristics are adduced, but frequently it is said that the situation is obvious.

8 It is generally recognized, however, that the wealthy dwellers in large homes are the largest domestic consumers of utility service.

9 *N. A. R. U. C. Proceedings,* XXXIX (1927), 448.

Nevertheless, the actual data on the income and wealth of domestic consumers in relation to the amount of gas or electricity consumed are fragmentary and scattered.

An often-cited analysis of consumer characteristics, now old, is that made by the Grand Rapids Gas Light Company of Grand Rapids, Michigan in 1922.[10] The company's cost analysis undertook to show that it was necessary for a customer to take at least 2700 cu. ft. per month in order to cover full costs of serving him. A sample group of customers consisting of widows, laborers, clerks, and other modestly paid workers used an average of slightly more than 2300 cu. ft. per month; this was almost enough to pay full costs of serving them. In contrast, bills for 1500 cu. ft per month or less almost always went to business establishments, professional men, childless middle-class families which ate meals outside the home or used prepared foods requiring little cooking, and well-to-do families which spent several months of the year away from home. This entire group constituted 30 per cent of all customers. No information was given on the type of customer who used more than 2700 cu. ft. From this data the company concluded that in Grand Rapids the consumers in modest circumstances used more gas than those apparently better off. The commission accepted this conclusion which can hardly be challenged.

The data in the Grand Rapids case have been used by advocates of the service charge as proof of the existence of similar conditions in other towns and cities. Additional information as to the relationship between consumption and economic status must be pieced together from secondary sources, or from data which permit inferences of limited validity. Henry G. Wells, a member of the Massachusetts Department of Public Utilities, once reported that the Department found that the class of gas

10 *Brief for the Grand Rapids Gas Light Co.*, filed by Travis, Merrick Warner, and Johnson in Case D 1649 before the Michigan Public Utilities Commission, February 5, 1923, pp. 17-27, 81-106. These data are summarized by W. G. Raymond, *op. cit.*, 125, and by the Commission in its opinion, *Re* Grand Rapids, P. U. R. 1923 C 453.

customers who consumed least were the so-called office consumers; the next smallest were merchants; the third lowest in consumption were churches, schools, and charitable institutions; the fourth were apartment dwellers; the fifth were residences where gas was used extensively; the largest consumers (except for large industrial consumers) were the ordinary household or domestic customers.[11] The latter customers lived for the most part in industrial centers, and were working people who used gas for cooking, washing, etc., largely as a matter of necessity.

In similar vein, the Massachusetts Department of Public Utilities stated in 1934:

> It has often been assumed by those who oppose a service charge that the small users of gas are the users of little or no means. Our investigations have indicated that this is not a fact. Persons of small means in the areas supplied by the Consolidated Gas Company [of Boston] are generally those who have a home life and substantial families and where a large amount of cooking is done and the use of hot water is substantial. We believe that a prohibition of the service charge will have little or no tendency to relieve the burden of those of limited means, but, on the other hand, will tend to increase them.[12]

This statement was based upon information derived largely from testimony presented to the Department of Public Utilities in rate proceedings, for the Department had never made any detailed survey of customer characteristics.[13] Evidence presented by the Boston Consolidated Gas Company in 1927

11 Henry G. Wells, testimony of April 17, 1925, United States Light and Heat Corporation *v.* Niagara Falls Gas and Electric Light Company et al., *Transcript of Record*, p. 272.

12 *Special Report of the Department of Public Utilities Relative to the Feasibility and Expediency of Prohibiting So-Called Service Charges in Domestic or Household Rate Schedules of Gas and Electric Companies* (D. P. U. 4693, Senate No. 50, 1934), p. 7.

13 Letter from Henry C. Attwill, then Chairman of the Massachusetts Department of Public Utilities, to the author, dated April 22, 1936.

undertook to show that, upon the basis of sample studies, gas consumption of wage earners was largest for those who lived in single-family houses, and that it declined as the number of families per dwelling increased to two, three, and four.[14] The small gas customers were principally those living in modern apartment houses, those who had servants and who used gas only to cook breakfast before the coal range was hot, and small stores and similar customers who still used gas for lighting.[15] Another study was made in 1929 for the same territory. It showed that a group of customers living in high-rental apartments consumed an average of 717 cu. ft. per month per customer, whereas the janitors living in the same apartment buildings, averaged 1731 cu. ft. per month. Moreover, customers who lived in single or small family dwellings in low rental districts consumed an average of 1774 cu. ft. per month.[16]

Evidence is united in showing that the ordinary apartment dweller generally consumes much less gas than the occupant of a single- or double-family house. Table 9 summarizes typical data on the consumption of gas in homes of various types, relative to single-family dwellings.

As the number of families occupying a building increases, the average consumption of gas per family decreases, though not in direct proportion. Additional information (summarized in Table 10 below) shows that apartment-house customers are relatively more frequent minimum bill customers (and hence small consumers) than are other customers.

Although apartment dwellers as a group have a lower average gas consumption than do other domestic customers, it is not

14 Dana D. Barnum, testimony on December 7, 1927, *Hearing on New Schedule of Rates Filed by Boston Consolidated Gas Co.*, Mass. D. P. U. 3031 (1927), p. 22.

15 *Ibid.*

16 Alan Cunningham, testimony on May 1, 1929, *Testimony of Witnesses for the Boston Consolidated Gas Company in the Matter of the Schedule of Rates Filed February 28, 1929*, Mass. D. P. U. 3530 (1929), Exhibit 11, pp. 66-70.

TABLE 9

RELATIVE CONSUMPTION OF GAS BY DOMESTIC CUSTOMERS ACCORDING TO TYPE OF DWELLING

Type of Dwelling	Relative Consumption per Family or Customer		
	General[a]	Northern Illinois[b]	Brooklyn[c]
One-family	100%	100%	100%
Two-family	68	69	79
Three-family	62	60	—
Four- to six-family	55	46	36
Over six families	46	—	—

[a] Data from estimates by Nash, *P. U. Rate Structures*, p. 180.

[b] Data pertain to the Public Service Company of Northern Illinois which serves suburbs of Chicago. Figures reported at first hand by W. J. Crowley and H. S. Jacobs, "Consumption Characteristics of Domestic Gas Customers," *Journal of Land and Public Utility Economics*, XII (1936), 127.

[c] Data pertain to the territory of the Brooklyn Borough Gas Company in the boroughs of Brooklyn and Queens, New York City. They are derived from an analysis of 1,000 customers of each type. Source: "A Study of the Consumption Characteristics of Customers Living in One- and Two-Family Houses and Apartments in the year 1928, together with Data on Real Estate Values and Rentals (Illustrated)," an unnumbered exhibit contained in the file of the record in Case 4458, Customers *v.* Brooklyn Borough Gas Co., New York Public Service Commission.

TABLE 10

MINIMUM BILL DATA ON DOMESTIC GAS CUSTOMERS IN NORTHERN ILLINOIS[a]

Type of Dwelling	Percentage of Minimum Bill Customers to Total Customers	Percentage of Minimum Bills Rendered to Total Bills Rendered	Average Number of Minimum Bills per Minimum Bill Customer per Annum
One-family	12.3%	3.6%	3.5
Two-family	25.2	7.6	3.6
Three-family	38.1	13.1	4.1
Four or more families	56.7	21.7	4.6
All dwellings	22.3	7.3	3.9

[a] Crowley and Jacobs, *op. cit.*, Table 1, p. 128. The survey showed that the number of minimum bills sent to apartment dwellers was absolutely as well as relatively greater than the number sent to single-family house customers.

apparent that all apartment customers are well-to-do.[17] There may be wide variations among apartment dwellers in gas consumption, and similar variations may occur within other classes. C. S. Reed declares, "Surveys in a number of places show that the higher the rent paid for an apartment, the less the gas bill." An exception occurs in the case of people who have apartments with a large staff of servants and who buy large quantities of gas. "As we descend in the scale of apartment rentals, there is an increased use of gas. . . . The less the rent, the greater the use of gas for both space and water heating."[18]

This statement is supported by data on the consumption of sample high-rental and low-rental apartment customers of the Brooklyn Union Gas Company in 1928 (see Table 11).[19] The high-rental group consisted of 459 apartment houses scattered through the territory; these contained 17,389 apartments of one to eleven rooms each, and rented for $12.00 to $90.00 per room per month. The low-rental group, located in the same territory, consisted of 172 apartment houses with 1234 apartments of one to seven rooms each, which rented for $3.75 to $9.00 per room per month. Although these two samples are not of equal size, certain rough conclusions seem tenable. The average monthly consumption per bill in the high-rental group was 1122 cu. ft., whereas in the low-rental group it was over three times as much, namely 3516 cu. ft. In the same year, the average monthly consumption for all bills rendered to customers was approximately 2700 cubic feet. According to the Company's analysis of costs, this brought in just enough revenue to cover the cost of rendering service plus a return on property investment. If the rate were adjusted to make all

17 See Table 13, *infra,* as to variations in monthly rentals of multi-family dwellings.

18 C. S. Reed, "Who the 'Small User' Really Is," *P. U. Fortnightly,* X (1932), 73.

19 The analysis is really of *bills rendered* rather than of customers. It is proper to say that 8.45% of bills in the high-rental group were sent out for no consumption at all, whereas it is highly improbable that so many *customers* would consume nothing all year round.

TABLE 11

COMPARISON OF GAS CONSUMPTION OF HIGH-RENTAL AND LOW-RENTAL APARTMENT CUSTOMERS OF THE BROOKLYN UNION GAS COMPANY, 1928 [a]

Monthly Consumption (cu. ft.)	Per cent of Bills Rendered High-Rental Group	Per cent of Bills Rendered Low-Rental Group	Per cent Accumulated High-Rental Group	Per cent Accumulated Low-Rental Group
000	8.5%	1.3%	8.5%	1.3%
200	8.4	1.3	16.9	2.6
400	9.9	1.8	26.8	4.4
600	10.9	2.3	37.7	6.7
800	10.7	3.2	48.4	9.9
1000	9.9	4.0	58.3	13.9
1200	8.6	4.4	66.9	18.3
1400	7.0	4.9	73.9	23.2
1600	5.6	5.4	79.5	28.6
1800	4.4	5.4	83.9	34.0
2000	3.4	4.8	87.3	38.8
2200–3000	8.7	20.2	96.0	59.0
3200–4000	2.7	13.0	98.7	72.0
4200–5000	.8	8.6	99.5	80.6
Over 5000	.5	19.4	100.0	100.0
Total	100.0%	100.0%		

[a] Data from Exhibits 40 and 59A submitted by the Brooklyn Union Gas Company, in Case 4832, New York Public Service Commission.

customers cover these costs (which are here assumed to be correct), about 96 per cent of the bills of high-rental apartment customers would have been increased, but only some 59 per cent of the low-rental group would have been raised. Furthermore, if a more moderate customer cost than the Company's figure of approximately $1.00 per customer per month were used, thus reducing the monthly average consumption necessary to recover full costs of service per customer, the comparison between the two groups of customers would be even more favorable to the low-rental group. Thus if 2000 cu. ft. were the point of equalization between rates and full costs of service, 83.9 per cent of the bills of the high-rental group would be increased by a readjustment of rates to recover full costs, but only 34 per cent of the bills of the low-rental group would be similarly raised.

From this evidence it appears that the apartment dweller with a small income would not be as adversely affected by the introduction of a service charge as would the customer living in the higher-rental apartment. This lends support to the argument for the service charge. However, similar data for other classes of dwellings are not available. It is highly probable that well-defined differences in consumption and income would be found among customers in detached homes. This assertion is suggested by the data on the variations in rentals of single-family dwellings shown in the following table.

TABLE 12

DISTRIBUTION OF GROSS MONTHLY RENTAL RATES, ALL RENTAL UNITS IN THE SIXTY-FOUR CITIES COVERED BY THE REAL PROPERTY INVENTORY OF 1934 [a]

Monthly Rental	All Dwelling Units	Single-family Dwelling Units	Multiple-family (All Other) Dwelling Units
Total number	1,597,208	676,341	920,687
Total—percentage	100.00%	100.00%	100.00%
Under $10.00	15.57	20.45	11.98
10 – 14.99	18.29	20.29	16.83
15 – 19.99	17.79	16.64	18.64
20 – 29.99	25.89	22.39	28.46
30 – 49.99	17.04	14.30	19.05
50 – 74.99	3.11	2.90	3.27
75 – 99.99	0.51	0.55	0.48
100 and over	0.30	0.38	0.23
Not reported and rent free	1.50	2.10	1.06

[a] Source: Stanley L. Kedzierski, *Markets for Electrical and Gas Appliances in Residences* (U. S. Bureau of Foreign and Domestic Commerce, Market Research Series No. 11: 1936, p. 104). Cities included in the Real Property Inventory of 1934 were distributed among all states, and contained 2,633,135 dwelling units of which 1,597,208 were classed as rental dwellings. Approximately 80 per cent of all *structures* surveyed were single-family units of five rooms, and 58.4 per cent of all dwelling quarters were in single-family buildings.

It is difficult to attempt to carry forward an analysis of subgroups in relationship to economic status because guiding data are almost totally lacking. However, it has been shown that there is a positive relationship between the number of rooms in a home and the consumption of gas by the occupant.[20] Since it can be also shown that generally the number of rooms in a dwelling unit increases as income increases (Table 13), there is at least a strong presumption that adequate data would indi-

TABLE 13

AVERAGE NUMBER OF ROOMS PER FAMILY DWELLING UNIT ACCORDING TO INCOME GROUPS FOR NINE U. S. CITIES, 1934 [a]

Income Group	Average Number of Rooms per Family: Austin, Texas	Birmingham, Alabama	Columbia, S. C.	Trenton, N. J.	Portland, Maine
All groups	4.9	4.9	4.8	6.2	6.0
No Income	5.1	4.4	4.3	5.8	5.3
$ 1–$499	4.0	3.9	3.6	5.9	5.4
500– 999	4.5	5.0	4.5	6.0	5.5
1000–1499	4.9	5.5	5.4	6.2	5.8
1500–1999	5.3	5.7	6.0	6.4	5.9
2000–2999	5.7	6.0	6.7	6.8	6.8
3000–4999	6.6	6.5	7.1	7.2	7.3
5000–6999	7.3	7.6	8.7	8.0	8.4
7000 and over ...	7.8	7.0	8.1	8.2	10.0

Income Group	Racine, Wis.	Fargo, N. D.	Salt Lake City, Utah	San Diego, Calif.
All groups	5.7	5.3	4.7	4.8
No Income	5.5	5.3	4.4	4.9
$ 1–$499	5.5	4.4	4.2	4.4
500– 999	5.7	4.9	4.5	4.4
1000–1499	5.8	5.0	4.6	4.7
1500–1999	5.7	5.3	4.9	5.0
2000–2999	6.2	6.0	5.2	5.2
3000–4999	6.5	6.6	6.1	5.9
5000–6999	6.8	7.8	6.5	6.4
7000 and over ...	8.4	8.7	7.4	7.7

[a] Source: Ada L. Bush, *Consumer Use of Selected Goods and Services* (U. S. Bureau of Foreign and Domestic Commerce, Market Research Series Nos. 5.1-5.9: 1935 and 1936).

cate that among single-family dwellers, low income would mean comparatively low consumption, and high income, large consumption. A similar presumption could be made for consumers living in apartment houses, since their consumption like-

20 Crowley and Jacobs, *op. cit.*, pp. 130-131.

wise seems to be related to the number of rooms occupied,[21] and since the size of quarters probably is related to family income. Yet this inference flies in the face of the chief conclusions from the evidence summarized above, namely, that low-rental apartment dwellers were larger consumers of gas than high-rental dwellers. This suggests that there probably is not as direct a relationship between the number of rooms occupied and family income for the apartment-dwelling consumers as a whole, as there is in the case of customers living in single-family houses. Data are not available to substantiate or disprove such speculation. At best, the evidence already summarized does not harmonize completely. Whatever conclusions can be drawn must be considered valid only with reference to the small and casual samples upon which they are based.

Characteristics of Electricity Consumers

Scanty though the data on gas consumers may be, comparable evidence as to electricity consumption is even scantier. A survey in 1923 of the two-year consumption records of electricity consumers in Massillon, Ohio, included (1) one hundred customers living in a shop and factory district, (2) one hundred customers living in a better residential district, (3) one hundred customers in a general residential and business district, (4) fifty residential customers in the steel mill district, (5) twenty apartment dwellers, and (6) fifty customers apiece in each of the three small communities served by the system.[22] In the supposedly better residential districts, the average consumption per customer per month was 22 kw.-hr., as against 24 kw.-hr. in the general residential and business district, 26 kw.-hr. in the factory district, 23 kw.-hr. in the steel mill section, and 24 kw.-hr. for the apartment-house customers. The depressed condition of business in 1921 and 1922 probably

21 Crowley and Jacobs, *op. cit.*, p. 131.

22 A. H. Heitzler and S. R. Finley, "An Analysis of Consumption and Revenue of Domestic and Commercial Lighting Customers," *N. E. L. A. Bulletin*, X (1923), 466-470.

affected the consumption of certain groups, especially those in the industrial districts. A graph of monthly data shows considerable crossing of consumption curves of the several groups, with one large peak for apartment-house customers. Hence the validity of the arithmetic mean seems doubtful. Even though variations from month to month are hidden, these data do not denote important differences in consumption characteristics between the supposedly well-to-do consumers (groups 2 and 5) and the poorer users (groups 1 and 4).

TABLE 14

PERCENTAGE DISTRIBUTION OF DOMESTIC CUSTOMERS OF THE PHILADELPHIA ELECTRIC COMPANY, ACCORDING TO ANNUAL CONSUMPTION, 1936 [a]

Annual Consumption	Percentage of Total
0–150 kw.-hr.	10.1
151–250 "	16.0
251–400 "	25.6
401–500 "	16.9
501–750 "	12.1
751–900 "	5.4
901–1200 "	7.5
over 1200 kw.-hr.	6.4
Total	100.0

[a] Monty, *op. cit.*, p. 3894.

More definite conclusions can be drawn from a survey of a large sample of " unprofitable customers " made by the Philadelphia Electric Company in 1936.[23] For the purposes of the survey, " unprofitable customers " were those who consumed 400 kw.-hr. or less per year. This amount was alleged to yield revenues barely sufficient to cover the costs of serving a customer. From the tabulation below, it appears that 51.7 per cent of domestic customers did not cover such costs (including capital charges). Moreover, 26.1 per cent did not pay their share of operating expenses. Examination of the occupations of

23 Austin Monty, " The Problem of the Unprofitable Customer," *Electrical World*, CVI (1936), 3893-3894, and " Selling the Unprofitable Customer," *Electrical World*, CVII (1937), 402-403.

customers using 400 kw.-hr. or less per year showed that a large group of them were unskilled workers; the remainder were skilled workers (employed in textile, garment, metal work, and building trades), small tradesmen, hucksters, shop clerks, unemployables, incompetents, or "professional poor." Especially outstanding is the fact that slightly more than one-half of these customers were employed; the remainder (43 per cent) were on relief. The annual earnings of those working ranged from $700 to $1800, and averaged $1100. Three-fourths of the entire group lived in brick-row houses with five or six rooms, and the remainder occupied two- or three-room dwellings. Generally, the electric wiring was in bad condition, and one-half of the homes did not have a single convenience outlet to facilitate appliance use. Lighting fixtures and lamps were of poor or sub-standard quality, and about one-half of all lamp sockets were empty. Though 94.0 per cent of the "unprofitable" customers possessed electric irons, 88.0 per cent radios, and 22.6 per cent washing machines, only 5.3 per cent owned electric refrigerators. In contrast with the relative scarcity of refrigerators among the "unprofitable" customers, 35 per cent of all domestic electric customers in Philadelphia had refrigerators, as did approximately 70 per cent of those customers who used more than 600 kw.-hr. a year. Clearly, the lack of money to spend on utility appliances and additional service constituted a definite barrier to increased consumption. The unprofitable customers in Philadelphia were not more well-to-do, or more able to pay a service charge, than those who consumed larger quantities of service.

Other data which bear upon the economic characteristics of various consumption groups are even more fragmentary than those detailed above. Yet they tend to support the conclusion that ability to pay, as indicated by income, is related to demand for service. A survey of the market for gas and electrical appliances in Appleton, Wisconsin, made by Time, Incorporated, revealed the information tabulated below. The fact that families receiving an income from $2,000 to $3,000 per

TABLE 15

ANNUAL AVERAGE CONSUMPTION OF GAS AND ELECTRICITY PER FAMILY IN APPLETON, WIS., 1932 [a]

Annual Family Income	Gas Cu. ft. per year	Electricity Kw.-hr. per year
Under $2,000	25,444	772
2,000– 2,999	22,343	741
3,000– 4,999	28,347	975
5,000– 9,999	33,146	1,045
10,000 and over	63,570	1,639

[a] Source: Time, Incorporated, *Markets by Incomes* (New York, 1932), II, 44.

year consumed on the average less the families in the lowest-income group supports the allegation that poorer customers use more service than do richer customers. This, however, does not account for the rise in consumption as incomes go beyond the $3,000 level. Unfortunately, the study of Appleton does not break down the broad lowest-income group into the sub-groups that undoubtedly exist.

Even among apartment dwellers who pursue similar occupations, there may be large variations in consumption. The following data (Table 16) relate to the outlays for gas and electricity made in 1931 by 400 families living in the Amalgamated Clothing Workers' Apartments in New York City. Unfortunately, the expenditures for gas are lumped together with those for electricity, and it is not possible to observe the relationship between the outlays for gas and those for electricity, and between each of these and family income. It cannot be assumed that the proportion of expenditures for gas to those for electricity remains the same throughout the entire range of income. With these reservations in mind, expenditures for utility service are closely related to income. Relatively few families which spent less than $2500 per year used mechanical refrigeration, even though they spent fairly large sums for ice. This study shows that among the clothing workers, the apartment dweller who was a small consumer of gas and electricity did not have a large income.

TABLE 16

EXPENDITURES OF FAMILIES OF CLOTHING WORKERS ON GAS AND ELECTRICITY, AND ICE [a]

Expenditure group	No. of families	Gas and Electricity [b]: Per cent of families purchasing	Gas and Electricity [b]: Average annual outlay per family	Ice [c]: Per cent of families purchasing	Ice [c]: Average annual outlay per family
$1000–$1499	17	100	$31.34	94.1	$25.18
1500– 1999	65	100	45.65	76.9	29.51
2000– 2499	89	100	50.34	73.0	30.92
2500– 2999	73	100	62.57	56.2	32.43
3000– 3499	55	100	64.87	54.5	35.58
3500– 3999	42	100	68.52	57.1	33.79
4000– 4499	26	100	76.81	50.0	30.28
4500– 4999	16	100	77.38	62.5	29.70
5000 and over	17	100	96.41	35.3	47.47
All families	400	100	59.69	63.3	31.65

[a] Data reported by A. Achinstein in *Report of the State Board of Housing on the Standard of Living of 400 Families in a Model Housing Project—The Amalgamated Housing Corporation* (Albany, N. Y.: Burland Printing Co., Inc., 1931), pp. 61-63.

[b] Gas and electricity is distributed as a co-operative enterprise; hence these expenditures are not directly comparable with payments to private utility companies.

[c] Mechanical refrigeration was used by 36.7 per cent of all families.

CONSUMER STATUS AND APPLIANCE POSSESSION

Data on the purchase or ownership of utility appliances shed an indirect light on the economic status of small customers. On a priori grounds alone, it may be said that poor customers may be small customers because they cannot afford to buy the more expensive appliances upon which large consumption must depend. It is not meant that there is a direct correlation between the purchase or ownership of an appliance and the consumption of service. Undoubtedly, there are large variations in the number of kw.-hr. consumed per kw. of maximum demand, and in the kw. of maximum demand per kw. of connected load.[24] Nevertheless, if consumers do not have appliances, their consumption is likely to be small.

An abundance of evidence shows that as the economic status of families improves, measured either by annual income or by rentals paid for housing, an increase takes place in the proportion of families which use electricity for lighting, and gas and electricity for cooking and mechanical refrigeration.[25] A striking fact, of no little significance, is that those tenants who pay $20 per month or less for rent, and who represent 51.65 per cent of all tenants in the sixty-four cities surveyed by the Real Property Inventory of 1934,[26] constitute a definite group in which the relative density of appliances is distinctly lower than in the group of tenants who pay high rentals. Even when the density of appliance use is measured in terms of those customers who use gas or electric service for some basic purpose, such as cooking or lighting, the application of utility service to a variety of purposes depends upon the family income. No general formula can be laid down, however, since there are wide variations in appliance density from city to city owing to climatic and economic forces. Apartments usually are better equipped for utilization of electricity and gas service, except possibly for water heating and house heating, than are single-family houses. This does not mean that apartment dwellers always, or even generally, consume more service than do single-family home dwellers. Little is known about how intensively each group utilizes appliances. But it does suggest that if the service-charge rate is to promote the sale of service, more appliances must be sold to customers with small incomes.

Summary

The answer to the question—who is the small consumer?—should be made only with reference to a given set of conditions.

24 See, for example, H. D. Rothwell, "Investigation of Domestic Demand," *The Bulletin of the Ontario Hydro-Electric Power Commission,* XVI (1929), 46.

25 This evidence is marshaled in Appendix D.

26 See, Table 12, *supra.*

Many well-to-do apartment dwellers are unremunerative consumers of gas, and, probably, of electricity. But often poorer families rely upon electricity and gas for essential household services and use service liberally. In some territories, however, the majority of customers are tenement dwellers whose low incomes permit them to use only whatever electricity is necessary for a minimum of lighting, and whatever gas is required for cooking. In such cases, a service charge burdens the poorer classes. Conditions of service and consumption vary considerably from territory to territory, and hence the effect of a particular service charge can be ascertained only by surveys in each area. The amount of the charge itself is an important factor in determining its social incidence. For, assuming that the smaller consumers are apartment dwellers, a large service charge affects the poorer strata more than a low one does. Almost any service-charge rate which increases the bills of one-half of all consumers must hit large numbers of low-income customers.

Should the Social Incidence of the Service Charge be Considered?

The discussion of economic characteristics of the small consumer of utility service presupposes that the social and economic incidence of a particular rate form should be given weight in designing a rate schedule. But rate-making sometimes seems to be a cold application of a mathematical cost theory. If such a procedure is valid, the effect of a rate upon the welfare of customers may be ignored, and the ability of the small consumer to pay his "share" of costs need not be considered. In contrast, it is often said to be proper to fix rates so as to socialize utility service. This may be done by selling service at less than cost to those who need it but cannot pay, and recovering losses from those who can afford to pay for more than their share of the cost. Under this price policy, the ability of the small consumer to pay is of vital importance.

Merely to state these views is to raise questions which transcend the bounds of these chapters. As a practical matter, no rate is a purely mathematical expression of a theoretical cost principle. Rates usually are a compromise between a theoretical rate determined by cost allocation and a rate that the customer is willing to pay. It is often said that the amount of the charge should not be fixed at more than the value of the service to the customer. Hence some speculation is necessary as to what the "value of the service" is for which the service charge is levied. This "value" cannot be fixed merely by the cost of alternative services which compete with gas or electricity. True, this might be a guide in rates for refrigeration, cooking, and water heating, for which the demand seems to be elastic. Small consumers, however, who use gas for cooking, or electricity mainly for lighting, would pay very high charges before they would stop taking service. Determined in this fashion, "value of service" would be equivalent to "all that the traffic would bear," and would represent not what the customers were *willing* to pay for the service without strong protest, but what they *had* to pay in order to obtain a necessary but monopolistic service. Therefore utility companies and commissions usually fix the service charge, not at some high figure based on a broad conception of customer costs, but at an amount trimmed to conform to customary charges which customers will pay without great objection. The "value of the service" in this sense is not determinable by a mathematical calculation, but only by a judgment made in view of rate history in a specific territory. The amount of the service charge which patrons are readily willing to pay, although not directly fixed by their incomes, is nevertheless strongly influenced by their economic status. In effect, therefore, the rate-maker must weigh the readiness of the small customer to pay, as well as the amount of customer costs, in setting the service charge. Regulatory commissions, as is shown in the following pages, give great weight to the objections of customers to the amount of the service charge. In doing so, they give effect to social considerations.

There are other grounds for recognizing the characteristics of small customers in making service charges. One is that it is desirable, as a matter of social policy, to obtain the widest extension of service, without, however, over-burdening the large users in consequence. Clearly this objective is forestalled if the service charge is so large that electric or gas service is prohibitively expensive to the many small users with low purchasing power.[27]

A social policy of rate-making may be supported on theoretical grounds. Where supply is rendered under conditions of increasing return, a subsidy may be used to increase the total economic utility accruing to the entire consumer body.[28] Where the total utility of the service is much greater than the total cost of rendering the service, as it is in the case of gas and electric domestic service, and where some of the consumers are unable to pay for the cost of the service, it may be considered a wise social policy to subsidize such customers in order to permit them to obtain the gratifications which they otherwise would lose. Such a subsidy might take the form of an abatement of taxes, but in this instance it takes the form of a subsidy to poor customers. The great obstacle in applying such a policy is that subjective utilities cannot be accurately determined. Hence, the application of a social theory of rate-making must rest largely upon the basis of observation of the role that electric or gas service plays in homes of the lower income classes.

27 This is suggested by Dr. E. W. Morehouse (Director, Division of Rates and Research, Public Service Commission of Wisconsin) in, "Supplementary Remarks on the Report of the Committee on Public Utility Rates," *N. A. R. U. C. Proceedings*, XLVII (1935), 205. It is significant that the Public Service Commission of Wisconsin follows a narrow interpretation of customer costs, and allows service charges of about $.60 for domestic electric customers. However, the Wisconsin Commission has not explicitly avowed the social policy mentioned above.

28 Alfred Marshall develops this point with full technical details in his *Principles of Economics* (Eighth Edition, London: The Macmillan Co., 1927), p. 469.

Despite the fact that a social policy may be desirable, it is clear that under present economic institutions, the subsidization of one class of consumers at the expense of another class cannot be carried beyond rather definite limits. By giving service at less than cost to small customers, a public utility would suffer a deficiency in revenues if it could not recover the loss from larger users owing to the fact that their patronage is competitive. Such a deficiency might impair the ability of the company to maintain the quality of its service, or expand and improve its service, to an extent which would far outweigh any benefits to the subsidized users. Furthermore, the losses on small customers would prevent a reduction in follow-on rates which might well increase output and thus improve utilization of plant and distribution facilities. This result is especially serious when the market for gas and electricity is an expanding one. For it is the expansion of service that may make possible the abolition of the service charge. A full development of the market for electricity or gas might well place a utility company in a financial position which would permit it to go without the service charge. What few small customers remained could easily be carried at a loss, especially if they were likely to take more service in due course of time. If the great majority of customers used large quantities of service, there would be little objection to a service charge buried in the first one or two blocks of the rate. The problem of the small consumer would disappear as the small consumer himself disappeared.

At present, the utility companies take the position that they should not be required to act as agencies for the redistribution of income by means of minimizing or abolishing the service charge. They hold that this task, if it is to be undertaken, should be performed directly by appropriate governmental bodies or by charitable organizations. The fact remains, however, that public utility companies, by their power to collect from consumers according to ability to pay for a necessary and monopolistic service, already exercise a governing power akin to that of taxation, which is a power to redistribute wealth. It would, therefore, be quite consistent with the quasi-public nature of public service corporations to require them to adjust rates for social purposes rather than merely for private ends.

Economic Effects of the Loss of Customers

The advantage of the service charge in recovering all or a part of customer costs from each customer is usually so great that companies often overlook collateral disadvantages. The service charge may cause some small customers to stop taking service. Moreover, high charges for small quantities of gas or electricity may discourage some people from applying for service. Each of these contingencies must be considered further.

By dropping small customers, the company reduces its total customer costs, but does not always save the full cost per customer. These customers are already attached to the lines of the company, and an investment has been made on each customer's premises and in distribution facilities. If the customer is lost, not all the costs theoretically allocable to him are avoided. Savings are only equal to the difference between the costs of having him connected to equipment already owned and the costs of not having him connected at all. There would be savings in commercial and maintenance expenses, but not in costs related to meter and service investment, unless the meters and services were later transferred to new customers. Whether such losses are sufficient to warrant discontinuation of the service charge depends upon the proportion of customers quitting service. Generally this is so small that the companies are willing to absorb the losses. Nevertheless, because of these reasons, one group of rate experts has reported that "in many situations even if it were possible to raise rates to small customers, it might be undesirable to do so." [29]

29 "Report of the Subcommittee on Water Heating Rates," *Report of the Rate Structure Committee of the American Gas Association, 1931* (New York, 1931), Appendix IV, p. 50.

The loss of some small customers is partly offset by the fact that others increase their consumption when a low energy or commodity rate accompanies the service charge. When this happens there is a two-fold gain. More customers use service liberally, and the company gets larger revenues which may exceed the losses from the defection of other customers.

The second effect of the service charge, namely, the discouragement of prospective patrons, appears in a somewhat different light. By foregoing the business of new customers, a public utility avoids the losses which would attend their attachment to the lines of the company. It is often maintained, however, that even though many prospective consumers would be carried at a loss for some time after they began to take service, they would steadily use more service as they became familiar with the benefits of gas and electricity, and so in time they would grow into profitable customers.[30] Moreover, if the service charge is prohibitive because it includes a large share of costs of distribution, it becomes necessary to load such costs on the existing large customers. If a fuller utilization of capacity could be attained by taking on small users at a rate to cover some small share of joint costs, the large customer would be relieved of at least that much of the joint cost. Accordingly, the Public Service Commission of Missouri points out that a heavy loading of costs in the service charge would make service too costly for many people, and concludes:

> Disregarding the social feature of the welfare of the individual being promoted by the general welfare of the community, there is a very real benefit to be derived by the large user of the commodity even though he pays part of the costs of the smaller user. The connection of many small users increases the load on the plant and gas utilities, among others, benefit by the economies of wholesale production. These economies are reflected in rates and the larger user may receive a net saving in his bill. . . .[31]

30 Watkins, *Electrical Rates,* pp. 69, 87; *In re* Electrical Companies (N. Y. P. S. C.), P. U. R. 1931 C 337.

31 The "costs" in question must include joint costs; otherwise it would

Though in many cases a present loss is avoided by not serving small customers who will not pay a service charge, a possible future gain will also be foregone. These gains are not susceptible to statistical measurement, and it is a matter of conjecture as to whether the prospect of future gains is certain enough to warrant present losses, even if the latter can be carried. The ultimate gains from taking on customers first served at a loss do not loom as large today as they did twenty years ago, when a much larger portion of the urban population was not served than now. But they do play a large part in the formulation of rate plans and financial policies for rural electric service, which remains largely undeveloped.

THE EFFECT OF OPPOSITION TO THE SERVICE CHARGE

The strongest practical argument against the service charge is that it is opposed by the public. The reasons for this opposition have been discussed above. Whether public hostility to the service charge is justified or not, the fact remains that it must be recognized. It adds to the feeling of the public against utilities in general and sometimes leads to legislative reprisals against utility companies. Equally serious is the effect it may have in creating a psychological barrier to the acceptance of promotional rates by the customer, thus negating a primary purpose of the service charge.

Rate-makers and regulators do not agree on the weight to be accorded to such antagonism. One school frowns upon the direct service charge, and prefers the minimum charge or some other type of indirect service charge.[32] In contrast, another group favors the direct service charge. It holds that popular opposition is overrated, and that it ceases when the service

not make sense to say that large consumers gain by paying a part of the costs of the small users. See Webb City *v.* Webb City and Carterville Gas Co. (Mo. P. S. C.), P. U. R. 1932 A 378. A similar view was expressed by the Missouri Commission in *Re* City of Carthage, P. U. R. 1932 A 389. See also *Re* Alabama Utilities Service Co. (Ala. P. S. C.), P. U. R. 1930 B 145.

32 See, for instance, the testimony of A. E. Forstall and V. M. F. Tallman, *Report of N. Y. Com. Rev. P. S. C. Law*, III, pp. 1919, 2279.

charge is properly explained to consumers and is given a fair chance to prove its merits in practice.[33] Instances are sometimes cited to show that public opposition changes to approval after the service charge has been in effect.[34] But men differ in their estimates of the strength of the opposition to the service charge in a given place. Clyde L. Seavey, when he was a member of the California Railroad Commission, concluded that the experience of utilities in California showed that generally, " . . . the service charge has been received without any but incidental protest which has been met by individual explanation." [35] On the other hand, D. F. Pegrum, after a careful study of rate-making in the same state, observed that utilities had met with considerable difficulty in educating the public to accept service-charge rates.[36]

The decisions and orders of public utility commissions disclose a similar difference of opinion among regulatory officials on the importance of antagonism to the service charge. In numerous cases commissions have refused to allow direct service charges in new schedules, or have required their elimination from old ones because of public hostility to them.[37] In other

33 "Report to the National Directing Committee of Executives," *Report of the Rate Structure Committee of the American Gas Association*, 1933, p. 37; C. S. Reed, "Who the 'Small User' Really Is," *P. U. Fortnightly*, X (1932), 78; W. J. Donovan, "Report of the Counsel," *Report of N. Y. Com. Rev. P. S. C. Law*, I, p. 114.

34 "Report of the Committee on Public Utility Rates," *N. A. R. U. C. Proceedings*, XXXI (1919), 68 (citing experience with state-wide gas service charge in New Hampshire); A. P. Elder and B. F. Bowers, *Transcript of Record*, U. S. Light and Heat Corporation *v.* The Niagara Falls Gas and Electric Light Company, et al., U. S. Circuit Court of Appeals, Second Circuit, 1930, pp. 390-396 (citing experience in Ottawa, Kansas, 1920-1924); *ibid.*, p. 481 (I. McDonnell as to experience in Montgomery, Alabama).

35 "The Disputed Service Charge," *P. U. Fortnightly*, VI (1930), 595.

36 D. F. Pegrum, *Rate Theories of the California Railroad Commission* (Berkeley, Calif.: University of California Press, 1932), p. 63; see also *Re* Pacific Gas and Electric Co., 1 P. U. R. (N. S.) 1 (1933).

37 The following cases are representative: *Re* Webb City and Cartersville Gas Co. (Mo. P. S. C.), P. U. R. 1922 C 608; St. Joseph *v.* St. Joseph Gas Co., 12 Mo. P. S. C. 142 (1922); *Re* Lansing Fuel and Gas Co. (Mich. P. S. C.), P. U. R. 1922 B 605; *Re* Fulton Fuel and Light Co. (N. Y. P. S. C., 2nd Dist.), P. U. R. 1921 C 787; *Re* Grand Rapids (Mich. P. S. C.),

cases, however, commissions have approved the use of the service charge despite vigorous protests against it.[38] These differences of opinion have become more and more important as state commissions have moved toward the establishment of standard rate types.[39]

In contrast with the experience in the United States, that in Germany indicates a ready acceptance of fixed charges by consumers. In the twenty-eight largest German cities of 200,000 population and over, according to Batson, two-part rates with a fixed meter charge prevail among the gas rates charged and are also common in electric rate schedules.[40] The failure of public opposition to submerge the service charge in Germany is especially interesting because the German gas and electricity enterprises, with a single exception, were owned by governmental bodies, or by mixed corporations whose shares were held by both public and private bodies.[41] Nevertheless, the dissatisfaction of small customers is the chief obstacle to fixed-charge rates even in Germany.[42]

It is difficult to give specific reasons for the startling contrast in American and German experience except by pointing out

P. U. R. 1923 C 453; *Re* New Jersey Gas Co. (N. J. B. P. U. C.), P. U. R. 1921 D 357; Borough of Wyoming *v.* Luzerne County Gas and Electric Co. (Pa. P. S. C.), P. U. R. 1923 D 80; *Re* Kansas City Gas Co. (Mo. P. S. C.), P. U. R. 1925 A 653; *In re* Electrical Companies (N. Y. P. S. C.), P. U. R. 1931 C 337; *Re* Portland Gas and Coke Co. (Ore. P. S. C.), 2 P. U. R. (N. S.) 277 (1934); *Re* Pacific Gas and Electric Co. (Calif. P. S. C.), 1 P. U. R. (N. S.) 1 (1933).

38 See, for example, Hartford *v.* Hartford Gas Light Co. (Conn. P. S. C.), P. U. R. 1920 F 840; Hyattsville *v.* Hyattsville Gas and Electric Co. (Md. P. U. C.), P. U. R. 1921 E 240; Keyport *v.* County Gas Co. (N. J. B. P. U. C.), P. U. R. 1928 C 325; Alabama Public Service Commssion *v.* Alabama Power Co., P. U. R. 1925 A 413.

39 See, pp. 144-5 *infra* as to the avoidance of public hostility to service charges by the adoption of initial charges.

40 H. E. Batson, *The Price Policies of German Public Utility Undertakings*, pp. 130-131.

41 Batson, *ibid.*, p. 56.

42 *Ibid.*, p. 134.

that meter charges in Germany were very modest, that they were graduated according to meter size, and that possibly they discriminated in favor of small customers. It may be that consumers served by public enterprises are less apt to criticize the form and amount of the charge in effect than are consumers served by private companies. But this supposition, even though it might be true of Germany, is one which the administrators of the Tennessee Valley Authority, and of the Rural Electrification Administration, have not tested in this country. Both these bodies have adopted as a standard rate form, a block rate with a simple minimum charge, and avoided a customer charge or initial charge.[43] Though the opposition of consumers to service charges may have caused these two public bodies to avoid a fixed charge of any kind, they have felt sure of a large consumption per customer, so that the minimum charge is sufficient protection against losses on small customers.

Alleviation of the Effects of the Service Charge

In order to avoid public hostility to the service charge, commissioners and utility executives sometimes have tried to introduce it in such a way as to alleviate its effects. One method has been to link it to a general reduction in rates. The general reduction somewhat offsets the new service charge, and a smaller number of bills are increased than there would be otherwise.[44] The company hopes that the new schedule will

43 The standard residential rate schedule enforced by the municipal and non-profit customers of the Tennessee Valley Authority, as of July 30, 1934, provides for a minimum monthly bill of $.75 per meter, which allows a use of 25 kw.-hr. The minimum charge for commercial customers is $1.00, which allows a use of 33 kw.-hr. per month. See, *Annual Report of the Tennessee Valley Authority for the Fiscal Year Ended June 30, 1934* (Washington, D. C., 1935), pp. 35-36. The Rural Electrification Administration advocates the use of a minimum charge varied according to operating conditions, and opposes any fixed charge for which no electricity is received by the customer. *Rural Electrification News*, I, No. 2 (October, 1935), 2, 28; *ibid.*, II, No. 3 (November, 1936), 2.

44 "Report of the Committee on Public Utility Rates," *N. A. R. U. C. Proceedings*, XXXVII (1925), 358; C. S. Reed, "Installing Better Forms of Rates for Gas," *A. G. A. Monthly*, IX (1927), 561-562.

stimulate consumption, so that total revenues will in time regain and perhaps exceed their old levels. Such action sacrifices present revenues, whether these be excessive or reasonable. The prospect of reduced revenues has deterred many utility companies from reducing rates to mitigate the effect of the service charge.

Optional Rates [45]

Another way to soften the opposition to the service-charge rate is to make it alternative to the old rate schedule.[46] In the late 1920's, so-called " optional rates " came into wide use in the electric light and power industry and, to lesser extent, in the gas industry. Their purpose was to promote business by permitting each customer to choose from several rates the one best suited to his requirements. In order that optional rates might attract patronage, they had to offer low follow-on energy rates, and these were most feasible in connection with a service charge of some kind. Also, the new service-charge rates were made optional so as to escape the opposition of customers whose bills would have been increased if the service charge had been levied upon all customers. Given a free choice, small customers remained on the old straight-line or block rate because the charges for small consumption were less than under

45 This subject cannot be treated exhaustively here. The emphasis is upon the relationship of optional rates to the service charge. The main sources are: Nash, *P. U. Rate Structures*, pp. 16-18, 162, 182-183, 315-317; Nash, *Economics of Public Utilities* (2nd edition, 1931), pp. 257, 281; F. A. Newton, " Competitive Rate-Making for Future Sales," *A. G. A. Proceedings*, XIII (1931), 405-407; "Report of the Committee on Public Utility Rates," *N. A. R. U. C. Proceedings*, XLIII (1931), 94-111; National Electric Light Association, *Report of the Committee on Promotional Rates* (1929), p. 4; *Report of the Rate Structure Committee of the American Gas Association* (1933), pp. 3-5, 40; *Re* Long Island Lighting Corporation (N. Y. P. S. C.), P. U. R. 1932 D 353; *Annual Report of the Department of Public Service, Public Service Commission, State of New York, 1931.*

46 The law of Mississippi now requires that gas, water, and electric companies which introduce service charges of any form, except flat demand rates, must offer an optional quantity rate not higher than the per-unit rates previously in effect. Mississippi Laws 1934, c. 318.

the new service-charge rate. In contrast, customers with a consumption larger than the "crossing point"[47] between the old and new rates, chose the new rate. The result of optional rates was that while charges for competitive uses were reduced, full costs were not recovered from small customers. The companies expected that the small customers would see the advantages of the low energy or commodity charges in the optional rate, and would transfer to it voluntarily. After some time, the proportion of customers remaining on the old rate would become so small that it could be abolished easily.

The spread of optional rates raised a new series of problems. As a device for effecting a painless transfer of customers from one rate to another, the optional service-charge rate succeeded only to the extent that it induced customers, especially small ones, to use more service. If most small customers failed to elect the new rate, the old rate could not be replaced by the new except by compulsion. This was a last resort because it raised the full force of customer opposition to the service charge. Yet if both schedules were continued, the problem of the small customer was not really solved. The losses on small customers had to be either absorbed by the company as a revenue cut, or loaded on the new follow-on rates. In the latter case, the new rate would be higher than would otherwise be necessary, and its power to stimulate consumption would be deadened. With several optional rates in effect, a desired amount of revenue can be collected only by making the level of one or more of them higher than it would be if it alone were in force.

A more direct cause of public dissatisfaction with optional rates has been the irritations attending their administration. With a free choice, customers might change frequently from one rate to another in accordance with monthly or seasonal changes in their consumption. This would increase commercial expenses, and would defeat the purpose of the optional rate. Hence, it has been required as a rule that the optional rate,

47 The point at which the bill for a given consumption is less under the new rate than under the old.

once chosen by the customer, applies for a definite period, ordinarily a year. In effect, the customer has to choose the rate that promises to be best in the future, even though he cannot foretell what his consumption will be. If he errs, the company gains. True, many utility companies have been willing to give the customer information to help him in selecting a rate, but the customer, not the company, has been responsible for the choice. The companies insist that they should not be expected to increase their commercial expenses by following the changes in the service requirements of each one of the thousands of customers on their books; nor should they be required to permit each customer to change from one rate to another whenever a change in consumption makes it profitable for him to do so.[48]

The policies followed by most companies in administering optional rates have led to objections by customers. They have not understood why they should not be placed automatically on the lowest rate, or why the company should not be required by the courts to make rebates when the bills paid under one rate are larger than those under an alternative rate.[49] Only lately have legislatures begun to take notice of the need for protecting the consumer. The chief example of legislation on optional rates is a Pennsylvania law enacted in 1937. This requires that where optional rates are in effect, the utility shall compute bills under the rate most advantageous to the patron.[50] Lacking legislative protection, large customers at times have protected themselves by paying free-lance rate ex-

48 Apparently there has been little uniformity among companies in advertising optional rates and in helping customers to make a good choice.

49 The courts have ruled that a public utility has no obligation to select the most favorable rate for the customer, nor to make reparation if an unfavorable rate has been applied. Silver's Lunch Stores, Inc. *v.* United Electric Light and Power Co. (N. Y. City Court, 1932), P. U. R. 1932 B 458; Spear & Co. *v.* Public Service Commission of Pennsylvania (Pa. Superior Court, 1932), P. U. R. 1932 D 382; Pantebakos *v.* Rockingham County Light and Power Co., 81 N. H. 441, 128 Atl. 534 (1925).

50 Pa. Laws 1937, c. 286, sec. 303.

perts a certain percentage of whatever savings they effect by discovering the most favorable rate. As a result some companies have had to choose between the alternatives of maintaining an advisory service of their own or of doing away with optional rates.[51] Some companies have found it better to make periodic audits to make sure that each customer is billed on the schedule most favorable to him, than to leave it to chance to show whether or not the customer has the best rate.[52] Despite these attempts to improve rate administration, some commissions are convinced of the fundamental weakness and inequity of optional rates. They are replacing them by a single rate for each class of customers, even though the change may increase some bills.[53] In the past, however, commissions as a whole have been inclined to favor optional rates because of their usefulness in introducing promotional rates.[54]

To summarize: the optional rate is useful in facilitating the introduction of a service charge rate. But the use of the optional rate is not certain to allay opposition to the service charge; it is difficult to administer, and it requires great care and diligence in order to avoid misunderstanding and opposition by consumers. The present movement for simplification of rate schedules portends an elimination of optional rates in favor of more simple rate structures.

51 P. M. Downing, "Rate Simplification as an Industry Need," *N. E. L. A. Proceedings*, XXCIX (1932), 25.

52 F. A. Newton, "Competitive Rate-Making for Future Sales," *A. G. A. Proceedings*, XIII (1931), 407.

53 Notably the New York Public Service Commission. See, *Annual Report of the Department of Public Service, Public Service Commission, New York* for the years 1931, 1932, 1934, 1935; *Re* Long Island Lighting Corp., P. U. R. 1932 D 353. See also, *Report of the South Carolina Power Investigating Committee* (Columbia, S. C., 1931), pp. 24-25, and *Re* Wisconsin Gas and Electric Co., 9 P. U. R. (N. S.) 517 (Wis. P. S. C., 1935).

54 "Report of Committee on Public Utility Rates," *N. A. R. U. C. Proceedings*, XLIII (1931), 97-99.

CHAPTER VI

ALTERNATIVES TO THE DIRECT SERVICE CHARGE

THE persistence of public opposition to the direct service or customer charge has made it expedient to use alternative methods of recovering customer and demand costs. These are now more common than the direct service charge.[1] The main substitutes are the minimum charge, the initial charge, and the block rate. They differ in the manner in which they are formulated and applied, and each, therefore, merits separate consideration.

THE MINIMUM CHARGE

The minimum charge, as ordinarily used, is a specific charge made in lieu of the charge for consumption when the latter (in dollars or units of service) falls below a specified amount.[2] Unlike the service charge, it disappears when the customer's consumption exceeds the minimum; in this respect it also differs from the initial charge. Ordinarily, the minimum charge for domestic service is uniform in amount, and is intended to cover at least some of the customer cost but not the demand cost. Sometimes, however, the minimum charge recovers demand costs, as is the case in rates for power or for special domestic loads such as electric cooking or house heating by gas.[3] Such

1 See pp. 35-39, *supra*.

2 The minimum charge is also used for purposes of rate differentiation, chiefly in power rate schedules. Difficulties in restricting the availability of special power rates sometimes are solved by specifying an appropriate minimum bill, or minimum consumption, or minimum demand, or some combination of these limitations. This prevents customers from obtaining a rate that is not intended to apply to them. See Nash, *P. U. Rate Schedules*, pp. 16, 84, 85; Eisenmenger, *Central Station Rates*, p. 98; Jones and Bigham, *Principles of Public Utilities*, p. 318; *Report of the Rate Research Committee*, N. E. L. A., 1931, p. 31.

3 Spurr, *Guiding Principles of Public Service Regulation*, III, pp. 397-398; H. Carl Wolf, "Electric Rate Structures," *Electrical World*, LXXXIX

a minimum charge may be varied according to some index of customer and demand costs, or it may be stated as a minimum consumption requirement, or as a definite minimum bill for a distinct type of service.[4] The problem of variable minimum charges is like that of variable service charges, which is discussed later in Chapter VII. The present chapter deals with the ordinary minimum charge.

If the minimum charge covers customer costs, it is improper to collect an additional service charge for that purpose. When commissions have found both a service charge and a minimum charge in one schedule, they usually have eliminated one or the other.[5] Yet in some cases the two may be properly combined. For example, the former may be used to require a consumption large enough to cover customer and demand costs which are not covered by a low service charge. This may be especially appropriate in appliance rates.[6]

(1927), 864; *Re* Princeton Electric Light Department (Wis. R. C.), P. U. R. 1927 E 437; Balaban and Katz Corp. *v.* Commonwealth Edison Co. (Ill. C. C.), P. U. R. 1929 D 305.

4 Though ordinarily the terms "minimum bill" and "minimum charge" are used as synonyms, they may pertain to different charges. Thus a rate schedule may specify a minimum charge for demand, and this may be only a part of the minimum bill for all service including a required amount of consumption. This distinction is of little importance here.

5 Spurr, *op. cit.*, III, p. 428; *Re* Enterprise Gas Company (N. J. B. P. U. C.), P. U. R. 1925 E 540; *Re* Fulton Fuel and Light Company (N. Y., 2nd Dist.), P. U. R. 1920 E 119; Fox *v.* Pine Grove Electric Light, Heat and Power Company (Pa. P. S. C.), P. U. R. 1920 B 380.

6 Domestic rates of the Ontario Hydro-Electric Power Commission contain a service charge plus an energy charge for kilowatt-hours used. If the sum of these is not at least equal to a minimum monthly charge, the latter prevails. (See, Ontario Hydro-Electric Power Commission, *Twenty-third Annual Report*, 1930 [Toronto, 1931], p. 386.) Some domestic rate schedules in Wisconsin make the customer charge equal to the minimum charge for demands of less than a certain standard, but impose a higher minimum charge (including the customer charge plus energy charge) when appliance capacity is more than a designated amount. (H. Zinder and W. E. Caine, "Regulation of Municipal Electric Utility Rates in Wisconsin: 1931-1932," *Journal of Land and Public Utility Economics*, IX [1935], 23.)

In effect, the ordinary minimum charge is a disguised service charge. The amount of the customer or demand costs hidden in the minimum charge may be approximated by subtracting from it the cost of the energy component that has been included. Consider, for example, the hidden customer costs in a schedule consisting of a minimum charge of $1.00 with an energy rate of seven cents per kw.-hr. If it is assumed that the energy cost is equivalent to the kw.-hr. rate, the customer cost collected from each patron would vary from $1.00, at no consumption, to two cents, at a consumption of fourteen kw.-hr. This estimate is, of course, inaccurate to the extent that the seven-cent rate includes customer costs, as it must in the present example. Otherwise, if all customers used more than fourteen kw.-hr. per month, the company would collect no customer costs at all. It is possible to make a closer approximation by applying, as the cost of energy, the unit price in that follow-on block in which customer costs are no longer a component. The demand element also could be segregated in a similar manner.[7]

No matter how they are ascertained, the imputed customer costs collected by a minimum charge vary inversely with the customer's consumption. There is discrimination among customers who pay a minimum charge as to the amount of customer costs they actually pay. Moreover, those consumers who use almost the minimum allowance, pay less customer costs than do those who take more than the minimum. The latter pay more than their share of customer costs in the energy or gas rate. And if the minimum charge includes all customer costs plus the cost of the average amount of electricity used by minimum-bill consumers, the very small customers pay their full share of customer costs in addition to a charge for allowed energy to which they are entitled but which they do not use.

When tested in this way, the minimum charge is discriminatory. It is much worse in this respect than an explicit service

7 Formulae for segregating the cost elements in a given rate schedule have been developed by H. E. Eisenmenger, *Central Station Rates*, pp. 118, 351-355.

charge, which is the same for all customers. But such discrimination is not objectionable if it is not excessive, and if it does not unduly burden the small customer whose consumption forms the basis for the minimum consumption allowance.[8] Whatever discrimination does occur is a necessary consequence of the attempt to hide customer costs. For these reasons, the discriminatory features of the minimum charge have been weighed lightly in balancing the relative merits of service and minimum charges. Nevertheless, in some cases they have helped to tip the beam in favor of the service charge.[9]

Promotional Aspects of the Minimum Charge

Does the minimum charge stimulate consumption? It is sometimes asserted that it induces small customers to use more gas or electricity. When they find that they regularly consume less than the amount allowed under the minimum charge, they increase their consumption in order to absorb the entire minimum allowance.[10] By this means, customers may become familiar with the benefits of a liberal use of gas or electricity, and may ultimately extend their consumption far beyond the limits of

8 See, for instance, *Re* Laclede Gas Light Company (Mo. P. S. C.), P. U. R. 1929 A 561.

9 "Report of the Rate Structure Committee," *A. G. A. Proceedings,* IV (General Session), (1922), 69; C. O. Ruggles, "The Significance of Minimum and Service Charges in Rate-Making for Public Utilities," *Journal of Political Economy,* XXXII (1924), 57; *Report of the Rate Structure Committee* (A. G. A., 1931), p. 31; Santa Barbara *v.* Southern Counties Gas Co. (Calif. R. C.), P. U. R. 1928 E 767; *Re* City of New London (Wis. P. S. C.), P. U. R. 1931 E 369; Phillips *v.* Boscobel (Wis. P. S. C.), P. U. R. 1932 B 17.

10 This has been said to lead to deliberate waste, especially when the minimum allowance is considerably above the level of consumption of small consumers. The New York Public Service Commission for the Second District held this to be an objectionable effect of a proposed increase of a minimum charge from $1.00 to $2.00 per month. (Town Board *v.* St. Lawrence Transmission Co., P. U. R. 1920 F 214.) But what constitutees "waste" of service must be a matter of opinion. It is conceivable that a community benefits from a minimum charge that causes customers to "waste" gas by using it to heat water for baths that they would not take otherwise.

the minimum allowance. Hence it is sometimes urged that the minimum charge is more promotional in effect than is the service charge.

Brief consideration shows that this argument lacks force. Although the minimum charge may induce the customer to increase his consumption without increasing his bill, it is difficult for him to gauge accurately the amount he has taken in a given period so as to know how much more to consume in order to reach the full minimum allowance. The mere fact that the minimum charge has been in effect for many years without actively stimulating consumption—as is unquestionably true in the gas industry—indicates that it is not promotional.[11] On the contrary, it may be said that the minimum charge is less promotional than is a fixed service charge. Because of the difference in the method of distributing customer costs, the quantity rate under the former must be higher for the first block, at least, than under the latter.[12] If the follow-on rate were the same in two schedules, one having a minimum charge and the other a service charge, and if it were intended to obtain equal revenues from both schedules, it would be necessary to make the minimum charge so much higher than the service charge that it might be prohibitive in effect.[13] If customers

11 Frequently, a substantial portion of revenues yielded by the minimum charge represents electricity or gas paid for but not used. Of the total minimum bill revenues collected by the Brooklyn Edison Company in 1929, approximately 36 per cent represented current paid for but not used; for the New York and Queens Electric Light and Power Company, the figure was 35 per cent. R. B. Grove, testimony in N. Y. P. S. C. Case No. 6367 (1930), *Stenographic Minutes*, p. 1814, and Exhibit No. 69 in the same case.

12 Watkins, *Electrical Rates*, p. 71; M. S. Sloan, testimony in N. Y. P. S. C. Case No. 6367 (1930), *Stenographic Minutes*, pp. 1364, 1512.

13 Evidence submitted in 1930 to the Public Service Commission of New York by four of the electric companies serving New York City bears on this point. It shows that revenues obtainable from a residential service charge of 60c per month with an energy rate of 5c per kw.-hr. could not be equaled by a minimum-charge rate with an energy charge also of 5c per kw.-hr., unless the minimum charge was $1.94 per month. N. Y. P. S. C., Case No. 6367 (1930), *Stenographic Minutes*, pp. 1241-1242, and Exhibits Nos. 45 and 54a.

refused to pay such a minimum charge, the quantity rate accompanying a more "reasonable" minimum charge would have to be considerably higher than under the service-charge schedule.[14]

From this it appears that the minimum charge is only slightly promotional, mainly within the limits of the minimum consumption amount.[15] The promotionality of a rate schedule, however, depends primarily upon the level of quantity rates, which may be low with either the minimum charge or the service charge if these cover sufficient customer costs. The minimum charge might best be used where the blocking of the follow-on rates induces a typical consumption high above the amount formerly allowed in the minimum bill. In such a case a service charge is not needed, and the minimum charge may be continued as a protective rather than as a promotional device.

Why the Minimum Charge is Used

The serious defects of the minimum charge seem to condemn its use as a means of recovering customer costs. Yet, even though these imperfections have long been evident, the minimum charge has prevailed over other types of customer charges because most consumers like it better than a fixed charge. Ratepayers are willing to pay for customer costs when they are sugar-coated by a small allowance of gas or electricity as they are in the minimum charge. If a minimum charge is in effect, each customer pays according to the amount of service registered on his meter. He believes, therefore, that the minimum charge is a payment for which gas or electricity is (or

14 To continue the illustration begun in the preceding footnote: if the minimum charge were reduced to $1.00, it would have been necessary to increase the energy rate to 6.4c per kw.-hr. C. S. Reed, testimony in N. Y. P. S. C. Case No. 6367 (1930), *Stenographic Minutes*, p. 1960.

15 R. B. Grove, testimony in N. Y. P. S. C. Case No. 6367 (1930), *Stenographic Minutes*, p. 1815. One observer believes that the result of the minimum bill is to remind customers to keep their bills low by not exceeding the minimum amount; H. Carl Wolf, "Electric Rate Structures," *Electrical World*, LXXXIX (1927), 864.

may be) received. Furthermore, because each consumer's requirements vary from month to month, relatively few customers fail to exceed the allowed minimum during some months of the year, and thus they avoid the minimum charge in those months. In contrast, the service charge recurs regularly each month, regardless of the amount of consumption.

For these reasons the minimum charge, unless excessive, does not arouse public antagonism. Hence public service commissions have often overlooked its disadvantages. They have been willing to regard the minimum charge as a fair compromise " which does not place upon small users the entire cost of serving them, but does require an appreciable contribution toward this end." [16] But the present acceptance of the minimum charge by the public is, in part, a result of long familiarity with it. When minimum charges were first introduced, they raised a storm of disapproval because they increased the bills of small consumers.[17] Perhaps, if present and future generations of consumers were to become familiar with the direct service charge, their objections to it might likewise disappear.

Monthly v. Annual Minimum Charge

No difficulties arise in choosing the billing period for the service charge: the variations in monthly consumption do not affect the twelve monthly payments. In the case of the minimum charge, however, if bills are rendered monthly a customer may pay a minimum bill for several months of the year even though his consumption during the remaining months may be large. His average monthly consumption may be higher than that of his neighbor whose consumption is steady but small. However, the latter never pays a minimum charge because he constantly uses an amount slightly in excess of the minimum allowance. Consequently, the first customer pays more per

16 *Re* Washington Heights Complaint (N. Y. P. S. C.), P. U. R. 1931 E 218.

17 Rochester Gas and Electric Corp. (N. Y. P. S. C., 2nd Dist.), P. U. R. 1921 A 415.

annum for a given amount of gas or electricity than the latter. In so doing, he contributes more toward customer and demand costs, and perhaps justly so, because he has a higher peak demand than the steady customer.

This apparent discrimination may be remedied by billing the minimum charge upon an annual instead of a monthly basis; a customer may then offset a small consumption in some months by a large use in others. An advantage of the annual basis is that it discourages customers who own summer cottages from having meters disconnected in order to avoid paying bills for winter months.[18] The annual basis is sometimes favored by law, as in Massachusetts, where a statute prohibits gas or electric companies from making a charge " for the use of a meter during any portion of twelve consecutive months if the consumer during that time uses electricity to the value of nine dollars, or gas to the value of seven dollars . . ." unless the approval of the Department of Public Utilities is secured.[19]

The annual basis, however, has weaknesses.[20] It is far less productive of revenue than the monthly basis. A company must carry a larger burden of customer costs in the energy or commodity rate if the minimum is applied annually, than if it is applied monthly. The annual minimum is not as useful as the monthly charge as a protection against losses on small customers. Moreover, with the case of a minimum bill accompanied by a block rate, the customer whose consumption is highly seasonal may pay the customer costs in the few months he takes service, but not in the other months. This happens because a large part of his consumption may be billed in the low-priced blocks that are not loaded with customer costs.

18 Another alternative, of course, is a substantial fee for turning the meters on or off.

19 Massachusetts General Laws (1932), c. 164, sec. 119; Acts and Resolves (1921), c. 404. With respect to gas, this prohibition goes back to Acts and Resolves (1886), c. 346, sec. 6.

20 Nash, *P. U. Rate Structures*, p. 299; "Report of the Rate Structure Committee," *A. G. A. Proceedings*, V (1923), 90.

The choice between the annual and the monthly minimum charge depends largely upon such factors as the homogeneity of customers and the blocking of the quantity rate. When the group to which the minimum charge applies consists chiefly of customers with similar characteristics and with regular and predictable consumption, it probably makes little difference whether the monthly or the annual basis is selected.[21] The monthly minimum certainly would be the simpler to apply.

The Initial Charge

In recent years, the initial charge has attracted more attention than any other alternative to the direct service charge. As used here, the term " initial charge " means a fixed charge for the first block of service, which is also a minimum bill. It is, for example, a monthly minimum charge of $1.00 for the first block of ten kw.-hr. (or any portion thereof) consumed per month. The remainder of the schedule usually consists of a blocked quantity charge.

The initial charge contains elements of the block rate and of the minimum charge. It is a flat charge for a very quickly breaking first block that is narrow compared with any follow-on block. There is a sharp contrast between the price per unit of service in the initial block, and the rate per unit in the ordinary follow-on blocks. The initial charge differs from the pure minimum charge. The latter loses its identity if a customer's consumption exceeds a specified amount; whereas the former

21 Watkins, *Electrical Rates*, p. 101. One proof of this is the fact that when each of the two bases is applied in comparable territories there is no substantial difference in the amount of current paid for but not used by minimum-bill customers. In 1929 approximately 36 per cent of the total revenues from minimum bills of the Brooklyn Edison Company, which made adjustments to an annual minimum, represented unused service allowance. In the same year, 35 per cent of the minimum-bill revenues of the New York and Queens Electric Light and Power Company, which enforced a monthly minimum without annual adjustments, represented current paid for but not used by customers. See, R. B. Grove, testimony in N. Y. P. S. C. Case 6367 (1930), *Stenographic Minutes*, p. 1814, and Exhibit 69 in the same case.

retains its identity. The purpose of making the initial block a minimum charge is to require customers to pay something even when they use no service at all.

The initial charge is a disguised or indirect service charge which, in common with the minimum charge, camouflages the customer-cost element with a small allowance of gas or electricity. It also discriminates among customers in collecting customer costs, but not as much as the minimum charge. The patron who takes the entire amount of, or more than, the initial block, still pays imputed customer costs; the rate per unit in the first block, if entirely consumed, is higher than the rate per unit in the succeeding block. In contrast, in the ordinary minimum charge, the rate per unit of the full minimum allowance is the same as the per unit charge for consumption immediately following the allowed minimum. This difference is illustrated by the following example.

	Rate A		Rate B	
	Initial charge $1 for first 10 kw.-hr.; excess energy, 5c per kw.-hr.		First block of 100 kw.-hr., 5c per kw.-hr.; minimum charge, $1	
Kw.-hr. consumed	Bill rendered	Imputed customer costs	Bill rendered	Imputed customer costs
0	$1.00	$1.00	$1.00	$1.00
5	1.00	.75	1.00	.75
10	1.00	.50	1.00	.50
20	1.50	.50	1.00	.00
30	2.00	.50	1.50	.00

These hypothetical rates do not pretend to yield the same revenues; but they do indicate the difference in the range of discrimination. The illustration shows that the initial charge recovers customer costs more uniformly than the plain minimum charge. The discriminatory features of the initial charge, like those of the minimum charge, must be regarded as a result of the attempt to avoid public opposition to a direct service charge.

The initial charge is likely to be more promotional than the minimum charge. Because the initial charge does not vanish, it is possible to collect the customer-cost component directly from each customer rather than largely in the quantity rate, as is necessary with the minimum charge. The initial charge differentiates the quantity rate from the customer-cost component more sharply than the minimum charge. This clear break makes it possible to emphasize the low quantity rate for all service consumed after the first block. For these reasons the initial charge is superior to the minimum bill.

The initial charge is superior to the direct service charge in some respects but inferior in others. It is more readily accepted by the rate-paying body than is the service charge; it is less effective in concentrating customer costs in a single charge. In order to cover a specific customer cost, an initial charge must be higher than a service charge based on the same cost. If the two charges are identical in amount, a larger part of the customer cost is recovered by the service charge than by the initial charge; this permits a lower follow-on rate than under the latter. Thus the initial charge carries the handicap of an energy or commodity cost. In practice, however, this disadvantage does not weigh heavily against it. An initial charge of $1.00 with a moderate amount of energy—e. g., 10 kw.-hr.—recovers a good part of customer costs; if these are placed at about 60 c. per customer, it will recover almost all of them. The fact that all of the customer costs are not collected in such an initial charge, especially if they are fairly large, is a recognition of the need for lightening the burden of the small customers. In this respect, the initial charge is a compromise with a charge that collects all of the customer costs from each customer.

The amount of gas or electricity covered by the initial charge is affected by a variety of factors. Often an attempt is made to fix it at the minimum amount which the customer is expected to require.[22] It may vary from one class of customers to another. Sometimes it is determined by the normal amount of

22 Nash, *P. U. Rate Structures*, p. 28.

service taken at the usual quantity rate.[23] The amount also may be found by cutting and trying, in an attempt to obtain the required revenues and, at the same time, to fix follow-on rates at a business-getting level. The furtherance of harmonious relations with the public,[24] and the necessity for meeting the requirements of laws that prohibit a direct service charge, also influence the amount of service allowance.[25]

The Initial Charge as an Alternative to an Illegal Service Charge

The initial charge is sometimes used to escape the reach of laws which prohibit the collection of meter rentals or service charges. The main instance of such a law is that of the State of New York which pertains to gas, but not to electricity.[26] The law provides that no gas corporation " shall make or impose an additional charge or fee for service or for the installation of apparatus or the use of apparatus installed." [27]

23 *Report of the Rate Structure Committee of the American Gas Association*, 1933, p. 38.

24 See, for instance, *Re* Alabama Utilities Service Company (Ala. P. S. C.), P. U. R. 1930 E 473.

25 See pp. 141-2 *infra*.

26 The restrictive laws in effect in Massachusetts before 1921 were strictly construed by the Department of Public Utilities to prohibit the use of direct service charges but not minimum charges or service charges concealed in block rates. (*Re* Customers of Boston Consolidated Gas Co., P. U. R. 1931 D 358.) In 1921 the Department was granted authority to allow service charges, and thereafter it favored direct service charges of moderate amounts rather than indirect charges. (Massachusetts Department of Public Utilities, *Special Report . . . Relative to the Feasibility and Expediency of Prohibiting So-called Service Charges . . .*, Senate No. 50, D. P. U. 4693, [1934].) The law of 1934 (Mass. Acts 1934, c. 365) restated the authority of the Department so as to allow it to make exceptions to a general prohibition of service charges for gas and electricity.

An Oklahoma law not only prohibits service charges in gas rates but also minimum charges; this makes it impossible to use the initial charge as a substitute for either. (Okla. Stat. 1931, sec. 11620; *Re* Avant Gas Service Co., 14 P. U. R. [N. S.] 391 [1936].)

27 N. Y. Public Service Law (1937), sec. 65 (6), Laws 1923, c. 898.

This provision was enacted in 1923 in order to halt a state-wide movement by gas companies to introduce service charges, which had been started after a holding by the New York Court of Appeals in 1922 that an old prohibition of gas meter rentals [28] could not be construed to prevent the collection of a direct service charge containing customer-cost elements.[29] The old law against meter rentals had been avoided by not counting the return and depreciation on meter investment as a customer cost.[30] This mode of legalizing service charges (which generally were fixed at about forty or fifty cents per month) could no longer be used after the explicit prohibition of gas service charges by the law of 1923. Hence gas companies resorted to the use of indirect service charges, ranging from 75c. to $1.00 per month, which were thinly veiled by the inclusion of 100 to 300 cubic feet of gas.[31] The Public Service Commission,[32] and the New York Supreme Court [33] both held that such rates complied with the letter of the 1923 law.

After a change in the chairmanship of the New York Commission, the law was construed differently.[34] In the Brooklyn

28 N. Y. Transportation Corporations Law (1937), sec. 16. This act is substantially a re-enactment of Laws 1868, c. 253, sec. 1.

29 Rochester *v.* Rochester Gas and Electric Corp., 233 N. Y. 39 (1922), 134 N. E. 828.

30 North Hempstead *v.* Public Service Corp. of Long Island, 199 App. Div. 189, 191 N. Y. S. 394 (2d Dep't. 1921). Also, the following cases of the Public Service Commission of the Second District, New York: Glove Workers' Union *v.* Fulton County Gas and Electric Co., P. U. R. 1921 C 789; Customers *v.* Cohoes Power and Light Corp., P. U. R. 1921 D 421; *Re* Rochester Gas and Electric Corp., P. U. R. 1921 A 415; Little Falls *v.* Utica Gas and Electric Co., 10 N. Y. P. S. C., Second District, 216 (1921); *Re* Kingston Gas and Electric Co., P. U. R. 1921 B 76.

31 In McCormick *v.* Westchester Lighting Company (N. Y. Sup. Ct.), P. U. R. 1931 E 6, evidence showed that in 37 localities in New York a charge of 50c to $1.60 per month was made for the first 100 cu. ft. of gas consumed.

32 Trustees *v.* Peekskill Lighting and Railroad Co., P. U. R. 1925 D 593.

33 McCormick *v.* Westchester Lighting Company, P. U. R. 1931 E 6.

34 Upon the resignation of Chairman William A. Prendergast in 1930, Dr. Milo R. Maltbie was appointed in his stead.

Borough and the Brooklyn Union Gas Company Cases,[35] the Commission held that the allowance of an amount so small as 200 cu. ft. in the initial charge had the effect of violating the spirit, if not the letter, of the law prohibiting service charges. Chairman Maltbie pointed out the absurdity of an initial block of 100 cu. ft. for $1.00 plus a commodity charge of 10c. per 100 cu. ft. If no minimum charge were collected,[36] this schedule resulted in a grossly disproportionate scale of charges; there was no charge for zero consumption, a charge of $1.00 for 100 cu. ft., and a charge of $1.10 for 200 cu. ft. A relatively small change in the consumption could not cause such a large change in the cost of serving the customer. In the opinion of the Commission, this defect could be remedied if the amount of gas included in the initial charge were substantial. Although the allowance might vary from company to company, it was thought reasonable if fixed in the neighborhood of 600 cu. ft. for an initial charge of $1.00.[37] The principles announced in the Brooklyn gas cases were later applied to electric as well as gas service charges in an effort to develop uniformity in rate types throughout the state.

The provision of the New York Law that forbids the collection of gas service charges should be repealed.[38] The prohi-

35 *Re* Brooklyn Borough Gas Co. and Brooklyn Union Gas Co., P. U. R. 1931 D 129 (Cases Nos. 4458 and 4832, N. Y. P. S. C.).

36 At the time there was no uniform practice in New York on this point. Some companies made no charge if no gas was used, whereas others levied a minimum charge.

37 Some later allowances were: 500 cu. ft. with an initial charge of $1.00 (Citizens' Committeee *v.* Kings County Lighting Co., P. U. R. 1933 B 151); 600 cu. ft. for $1.10 (*Re* Republic Light, Heat and Power Co., Inc., P. U. R. 1932 B 512).

In contrast to the view of the New York Commission, the Massachusetts Department of Public Utilities has taken a narrow, legalistic position, and has held that an initial charge of sixty cents for 100 cu. ft. of gas per month does not violate the terms of a statute prohibiting service charges. *Re* Boston Consolidated Gas Co., 12 P. U. R. (N. S.) 113 (1936).

38 The constitutional validity of a service-charge prohibition is not settled, The argument against such a statute is that a company is legally obligated

bition is at present unnecessary because the Public Service Commission favors the use of initial charges on the ground that they are less objectionable to consumers than direct service charges. Even if the prohibition were effective, it would be undesirable in itself, for it would prevent the orderly development of rate differentiation. It is noteworthy that repeal was recommended by the New York Commission on the Revision of the Public Service Commissions Law in 1930, with the proviso that service charges should be based upon careful analyses of costs of service and should not be larger than the out-of-pocket customer costs.[39]

to furnish facilities on the customer's premises; hence, where little or no service is used during the month and no service charge is collected, the compensation is inadequate and the company is deprived of its property without due process of law in violation of the Fourteenth Amendment of the Constitution of the United States. As yet no court has settled the question on its merits. In 1922 the New York Court of Appeals held that a service charge was not prohibited by the interdict against meter rentals, but it did not rule on the legality of a direct service charge prohibition such as was enacted in 1923. (City of Rochester *v.* Rochester Gas and Electric Corp., 233 N. Y. 39, 134 N. E. 828 [1922].) In New York and Richmond Gas Co. *v.* Prendergast, 10 Fed. (2d) 167 (E. D. N. Y., 1925), P. U. R. 1925 E 21, it was held that a specific rate which was limited by a maximum rate statute, as well as by a service-charge prohibition, was confiscatory. This decision, however, did not void the service-charge statute standing alone.

The question of constitutionality was most urgently raised in United States Light and Heat Corp. *v.* Niagara Falls Gas and Electric Light Co., 23 Fed. (2d) 719 (W. D. N. Y., 1927). The court held that the New York statute forbidding a gas service charge was unconstitutional: (1) with reference to the petitioner, who was a prospective customer of the defendant, for his property right to be served at a reasonable rate would be denied if a service charge were not made a part of the rate schedule in order to allot costs among customers equitably; (2) with reference to the defendant since the law denied the right to earn a reasonable rate and thereby compelled the defendant to serve at a yearly loss. This decision was reversed by the Circuit Court of Appeals, 47 Fed. (2d) 567 (C. C. A. 2d, 1931), on the ground that a customer had no property right which was violated by the statute prohibiting a service charge. Certiorari was denied by the United States Supreme Court, 283 U. S. 864 (1931).

39 *Report of N. Y. Com. Rev. P. S. C. Law*, I, pp. 115, 418.

The Initial Charge v. the Direct Service Charge

The main objection to the initial charge is that it does not require the rate-paying public to pay for customer costs directly and openly.[40] It has already been shown that this charge is discriminatory and less promotional than a direct service charge. Yet most rate-makers believe that customer costs cannot be assessed openly and directly without a great strain upon the relations between utility companies and consumers.[41] The initial charge, therefore, is favored as a compromise with the hope that it may prove to be a stepping stone to the direct service charge.

The sharp difference of opinion on whether the service charge should be open or concealed is illustrated in the report of the Committee on Public Utility Rates of the National Association of Railroad and Utilities Commissioners in 1935.[42] The committee recommended that all companies should adopt a uniform type of rate, namely, a block meter rate that would include an indirect service charge and that would be priced and blocked to develop the heavy appliance load.[43] The committee declared that the initial charge is superior to the service charge from the standpoint of consumer relations and increased sales, because it does not erect a psychological barrier to liberal consumption. It cited with approval the use of the initial charge by the gas and electrical utilities in New York as a result of

40 Watkins, *Electrical Rates*, p. 100; F. T. Singleton (member of the Indiana Public Service Commission), testimony, *Report of N. Y. Com. Rev. P. S. C. Law*, II, 1516-1518; F. P. Morgan, "Discussion of the Report of the Committee on Public Utility Rates," *N. A. R. U. C. Proceedings*, XLI (1929), 441-444.

41 See, pp. 121-3, *supra*; also "Report of the Committee on Public Utility Rates," *N. A. R. U. C. Proceedings*, XLI (1929), 337-339, and XLVII (1935), 195-204.

42 "Uniformity of Electric Rate Forms and Simplification of Electric Rates," *N. A. R. U. C. Proceedings*, XLVII (1935), 195-206, 449-454.

43 *Ibid.*, p. 201.

the leadership of the Public Service Commission of New York.[44]

A different opinion, favoring the direct service charge, was expressed by one member of the committee, Dr. E. W. Morehouse, who unofficially voiced the view of the Public Service Commission of Wisconsin. He considered it unnecessary to practice deception in assessing readily known customer costs when frankness is justified. Though he was fully aware of the weight of public opposition to the service charge, he urged that a policy of openness should first be followed. The Wisconsin Commission has followed this policy in recent years. In 1931 it set out to simplify and standardize rates for household and small-store customers, and wherever possible required utilities to use a service charge plus block-energy type of rate. As a result, this type has displaced other and often more complicated schedules; by 1935 about three-fourths of the urban population of Wisconsin was billed under the service-charge rate for household and store uses.[45] Thus far, the service-charge rate adopted in Wisconsin has been satisfactory. A large part of its success however, can be traced to the fact that the service charges collected are moderate in amount; usually they are sixty cents for residential service and seventy-five cents for commercial service. They cover only those average, actual, and directly traceable customer costs which do not vary materially between individual customers in the two classes of service. The Wisconsin Commission does not approve of service charges of $1.00 a month or more, since it feels that the complaints against the service charge usually arise because of its excessive amount.[46]

44 In the period from 1930 through 1935, the New York Commission eliminated service charges in 119 schedules. *Annual Report of the Department of Service, Public Service Commission*, 1935, p. 18.

45 E. W. Morehouse, "Supplementary Remarks on the Report of the Committee on Public Utility Rates," *N. A. R. U. C. Proceedings*, XLVII (1935), 205. See also, *Biennial Report of the Public Service Commission of Wisconsin*, 1932-1934 (Madison, Wis., 1934), pp. 12-14.

46 E. W. Morehouse, *op. cit.*, pp. 449, 450, 461. Other important reasons

Other Alternatives to the Service Charge

The only remaining practical alternative to the direct or indirect service charge is the block rate. Straight-line or step rates are entirely unsuited to the recovery of customer or demand costs for most companies, though they might be where there was almost absolute uniformity in the amount, time, and steadiness of consumption and demand of all customers. Flat demand rates, on the other extreme, do not recognize variations in consumption. They should not be used except when definite limitations may be readily imposed upon the amount of consumption and demand. The outstanding advantage of flat demand rates for small customers is that no meters are needed. If such rates were put into effect, there would be a great saving in customer costs simply by wiping out carrying charges on meter investment and reducing commercial expense.[47] Charges to small customers could be made much lower for a given quantity of gas or electricity than under a meter rate, and losses on customer costs would be minimized.

Would customers be receptive to a bill consisting entirely of a fixed charge, if they object merely to a customer charge? Prospective customers might favor the fixed charge largely because it would enable them to ascertain their monthly bill in advance. The question of what a new service would cost could be answered definitely. But old customers would find that their bills would grow larger as the introduction of new appliances increased their demands. Experience with flat rate schemes in which the customer's individual demand was controlled by a demand-limiting device, has shown that such rates ultimately prove irksome to customers whose demands fluctuate or grow as time goes on. They usually prove to be more troublesome to

for the success of the service-charge rate in Wisconsin are that it replaced a variety of complicated schedules such as the room- or connected-load types of rates which irritated customers, and that it was accompanied by general rate reductions.

47 Eisenmenger, *Central Station Rates*, p. 128.

administer than had been anticipated.[48] It is feared, moreover, that if customers are billed a flat demand rate, they will consume gas or electricity for frivolous purposes and thus "waste" service. But in some instances the cost of gas or electricity "wasted" under a flat rate is so low that it is more than balanced by the savings in customer costs.[49] Nevertheless, utility managers in the United States have avoided flat demand rates, except for specified and separable loans, e. g. water heating, because of their fear of service wastage and of the difficulties of controlling individual demands. Moreover, flat demand rates tend to cause customers to restrict the installation of appliances in order to avoid larger bills. Therefore, they are not used to any great extent for domestic service in the gas and electricity supply industries. There have been but few attempts to apply them to small domestic customers as a group, since the days when high-efficiency lamps were first introduced.[50]

The Block Rate

The block meter rate per se may be used for the recovery of customer and demand costs if the width and height of each block are carefully determined. It has the advantage of being purely a quantity rate, and escapes the opprobrium of a fixed-charge rate. The block rate, however, fails entirely to recover customer or demand costs from those customers who use no service at all. To recover customer costs from small customers it is necessary to employ a short first block, which, if priced high enough, is the equivalent of an initial charge. Used in this way, with a minimum charge, the block rate is an indirect service charge. When the rate is not blocked to include the

48 Spurr, *Guiding Principles of Public Service Regulation*, III, p. 368.

49 Eisenmenger, *op. cit.*, p. 199, n. 1. This is less true of gas than electricity for usually the commodity costs of gas are a larger part of the total costs of service than they are in the case of electricity supply.

50 The use of flat demand rates for small residential customers in some sections of the country is cited in the *Report of the Promotional Rates Committee*, National Electric Light Association, July, 1929 (Publication N. 289-93), p. 2.

equivalent of a fixed charge, because of the desire to keep the top rates for gas or energy at a low level, it frequently happens that the first block is so wide that it is in effect a straight-line rate for most residential customers.[51] Some companies, however, do use a block rate successfully. They offer customers the possibility of reaching a low unit rate in those blocks corresponding to the service used in operating the larger domestic appliances. Though they retain the minimum charge as a protective device, they rely upon proper blocking to solve the problem of the small customer by inducing him to become a large customer. In short, the success of the block rate in recovering customer costs depends largely upon the formation of the blocks in order to collect the equivalent of a fixed charge from all customers.

Another type of schedule which might be used to give proper recognition to customer and demand costs is one which increases or decreases the unit charge as consumption falls or rises, with a service charge applying only when no service is taken. Such a rate, however, though it might be equitable, would be so complicated and cumbersome, so subject to mistakes, and so expensive to administer that it cannot be considered practicable.[52]

Summary

The use of indirect forms of the service charge rests upon the need for recognizing the public opposition to the service charge in its direct form. The initial charge is superior to a minimum charge in that it is less discriminatory and more promotional than is the ordinary minimum charge by itself. As a

51 W. J. Donovan, as counsel to the New York Commission on the Revision of Public Service Commissions Law, said of the domestic gas block rates prevailing in 1930 that " in most of the cases the first block is so large that it is rarely exceeded by the ordinary domestic consumer so that in effect the rate becomes a straight line charge," *Report*, I, p. 113. See, also, *Report of the Rate Structure Committee*, A. G. A., 1933, p. 6.

52 C. L. Seavey, " The Disputed Service Charge," *P. U. Fortnightly*, VI (1930), 594.

protective device, the minimum charge might well be employed with a block rate which, by offering low increment rates, raises domestic consumption to high levels. The present trend, however, is toward the use of the initial charge followed by a block rate of two or three blocks which are adjusted to obtain the loads of large appliances.

CHAPTER VII

VARIABLE OR UNIFORM SERVICE CHARGES?

As a rule the service charge is levied as a uniform charge per meter, or per customer. But some cost analyses logically lead to a service charge graded according to variations in customer or demand costs from one customer, or group of customers, to another. The basis of the variation depends upon the analyst's conception of the costs that should enter the service charge.

Both elements of customer cost—i. e., operating expenses and capital charges—may be shown to vary between groups of consumers. By and large, however, operating expenses are taken to be approximately equal for all customers, whether large or small. Can such expenses be considered uniform per customer? Certain customer operating expenses, such as distribution superintendence and labor, depreciation, taxes, etc., which are apportioned as functions of capital investment, must vary among customers if the investments on customers' premises are different. It is probable, moreover, that certain of these expenses, such as distribution superintendence and labor, are lower per meter of a given size for meters in apartment dwellings than for meters of the same size located in scattered single family dwellings. But there seems to be no practicable way of tracing these costs on the basis of density of customer population,[1] and hence averages must be used.

There remain for consideration those customer operating expenses (consisting of commercial, utilization, new business, meter setting and removing, and general expenses) that are not a function of investment. If the commercial expenses of dealing with special contract or large power customers are eliminated, the remaining customer operating expenses would

1 John E. Godsoe, testimony in *In re* Brooklyn Gas Co., N. Y. P. S. C. Case No. 4458 (On Rehearing), *Stenographic Minutes*, pp. 1638-1678.

seem to be allocable to all customers equally. But even on a priori grounds, it would seem that there is some variation in such expenses per meter.[2] The fragmentary evidence that is available points to the conclusion that the number of meters read in one day by meter readers, or the number of bills delivered and collected by hand, depends largely upon whether the district covered is a densely populated apartment area, or a district of one family houses.[3] The cost of reading commercial meters is probably less than for residential meters, since there are fewer occasions for two or three visits by a meter reader to obtain a reading.[4] But such variations in commercial expenses usually are not accurately ascertained. This is also true of utilization, new business, and meter maintenance expenses. Ordinarily, it is assumed that if variations from the average were isolated, they would be negligible. Furthermore, any attempt to recognize them would lead to confusion and complexity in rates, which probably would increase the expense of administration more than enough to offset the slight gain in wiping out the discrimination from the process of averaging these costs. And finally, as a practical matter of administration such variations are difficult to express in rate schedules. In practically all calculations of customer costs, therefore, customer operating expenses are averaged for the entire group of customers. At present little is actually known as to whether or not the variations in these costs are so small that they may be ignored without discriminating against some customers unduly.

The major reason for graduating a service charge is that the investment attributable to the customer per se may vary from one customer to another. Since there are a variety of

2 Eisenmenger, *Central Station Rates*, pp. 48, 49.

3 W. O. Mulligan, "Domestic Meter Reading and Bill Delivery," *Pacific Coast Gas Association Proceedings*, XXXII (1932), 63-70; G. B. Webber, "Measuring Meter Reading Time and Requirements," *A. G. A. Monthly*, XVI (1934), 129; Samuel Newman, *In re* Brooklyn Gas Co., N. Y. P. S. C. Case 4458 (On Rehearing), *Stenographic Minutes*, pp. 1356-1359.

4 H. B. Gear, discussion of paper by N. R. Gibson, *Electrical World*, CI (1933), 751.

views as to what constitutes customer investment, there are likewise several views on the ascertainment and application of the graded service charge. The different types of variable or uniform service charges are considered in turn below. The equitableness of each kind of service charge can only be determined by considering it in relation to the remainder of the schedule.

Theoretical Basis for a Uniform Service Charge

A uniform service charge may be justifiable if it is based upon uniform elements of customer costs. A correct non-variable customer cost may be determined by considering customer investment to be solely the investment in meters and service lines or pipes. Up to a certain size of customer, such investment is practically constant. Thus, the customer investment may be divided into two parts: (a) the constant part, which is the cost of the meter of minimum size, plus the cost of a service connection of average length and minimum capacity; (b) the variable part, which is the remaining investment in the meter and the service connection, and is proportional to the maximum demand of the customer.[5] If capital charges on the variable investment are assigned to demand costs, the charges on the constant investment are independent of the size of the customer. The latter, together with customer operating expenses, form the customer cost which is constant per customer.[6]

It follows that a service charge which covers customer costs viewed narrowly—as above—should not be graded. Those costs pertaining to the variable elements in the meter and service investment should be covered in the demand and consumption

5 Eisenmenger, *Central Station Rates*, pp. 46-47. A substantially similar view was taken by Forstall who, however, mentioned the average size of meter instead of the minimum size as the dividing line between constant and variable customer investment; A. E. Forstall, "Methods of Charging for Gas," *American Gas Light Association Proceedings*, XXII (1905), lxxxiv-lxxxix. See also Harry Barker, *Public Utility Rates* (1917), p. 287.

6 It is assumed, of course, that customer operating expenses are uniform, owing to the difficulties of cost apportionment.

components of the rate. A uniform service charge of this type does not cover all costs traceable directly to customers, unless all consumers are served by meters and services of one size. Under ordinary conditions, therefore, a uniform service charge governed by the investment in meters and services of minimum size would be less promotional than other uniform service charges. For, without a separate demand charge, large customers would have to bear the variable meter and service line costs in the form of energy charges higher than would be necessary with a larger service charge. Yet, if a uniform service charge were made to cover all customer costs, averaged for all customers, it would unduly burden small customers, except in areas in which all meters and services were the same in size. This type of uniform service charge increases the difficulties of introducing multi-part promotional schedules. Of course, if the consumer population were of a homogeneous character it would make little or no difference whether one approach or the other to the uniform service charge were used. Finally, the service charge might best be made uniform merely because customers might fail to understand a graded charge, and would more readily accept some moderate uniform charge as being fair.

Customer Charge Based Upon Meter Capacity

It has been held by some writers that the variable service and meter investment, treated above as a demand element, ought to be included in the service charge. Watkins suggests the use of a "meter charge" which covers, (1) the constant elements of operating cost and capital charges on meter investment, and (2) the variable capital charges arising out of the fact that a customer with a heavy demand uses a meter much larger than one of minimum capacity.[7] The meter charge would be graduated upward as the meter costs mounted upward.

7 G. P. Watkins, *Electrical Rates*, pp. 68-70, 97-99. The term "meter charge" is distinguished from the "consumer charge" which is a non-variable charge per customer.

Watkins excludes service line or pipe from customer costs. He does this on the ground that, in many urban localities, the service line is jointly used, and is a part of the distribution line: it is not a part of the direct separable costs traceable to the fact of a customer's connection.[8] Other writers, however, have included the service line in a variable customer charge, and it is general practice to do so, despite the fact that the investment in such lines is not the same for all customers. In an early discussion it was pointed out that items such as maintenance and interest on meters and service pipes should vary with the sizes of each and with the length of service pipe required.[9] A further refinement would be to give weight to the number of meters per service in order to give apartment dwellers their due. But in practice it is too difficult to administer a charge graded according to four factors—meter capacity, service line capacity, length of service line or pipe, and number of meters per service.

In contrast to the uniform consumer charge, Watkins' meter charge would provide for some degree of graduation according to the number of meters a consumer might have,[10] and also according to the size of the meter. It would recognize the demand elements which are ruled out of the pure customer charge as conceived by Eisenmenger. If a demand charge is to be a part of the rate structure, the meter charge should be carefully formulated so as not to duplicate some demand elements. If, however, as generally is the case, an explicit demand charge is not a part of the schedule, then, it is argued by Watkins, the meter charge would serve in part as a demand charge.

The basis of the meter charge is the prime cost of the service to the individual exclusive of energy demanded and consumed.

8 *Ibid.*, p. 89.

9 Walton Clark, "Meter Rents—A Question of Equity and Policy," *Proceedings of the American Gas Light Association*, IX (1891), 522.

10 This does not radically depart from ordinary practice in which customer charges are levied on a per meter basis rather than per customer.

This differs from the basis of the demand charge, which is the fixed cost imposed upon the company by the customer's demand. Although the meter charge does cover some demand costs, it does not explicitly give weight to load factor considerations. Yet, when meter sizes are carefully adjusted to the needs of the customer, the meter capacity (in watts) is more directly related to maximum demand than to connected load, for it need not be adjusted to aggregate connected capacity, but to the prospective maximum demand. Thus the meter charge, reflecting variations in meter capacity, gives effect to demand cost elements related to the meter.

The meter charge has the advantage of being more simply and easily administered than most demand charges. Despite this advantage, service charges varying according to meter capacity, though not uncommon, are not the general rule. In most instances, the grading of a service charge would not materially affect payments made by a majority of customers because rates for exceptionally large consumers are stated as class rates, and because few domestic patrons require large capacity meters.[11] Nevertheless, public service commissions sometimes approve or require the use of customer charges varied according to meter capacity. In many of these instances the classification of charges and meter capacities is elementary.[12] In other cases, however, service charges on a meter basis have been disapproved as improperly recognizing something of a demand element, which, if not segregated by itself in the rate, would be more properly absorbed in the consumption charge.[13]

11 Schaub *v.* Mechanicsburg Gas and Water Co. (Pa. P. S. C.), P. U. R. 1920 B 258.

12 *Re* Malden and Melrose Gas Light Co. (Mass. D. P. U.), P. U. R. 1922 C 545; *Re* Malden Electric Co. (Mass. D. P. U.), P. U. R. 1922 C 726; *Re* New Jersey Gas Co. (N. J. B. P. U.), P. U. R. 1918 B 438; *Re* Georgia Power Co. (Ga. P. S. C.), P. U. R. 1929 B 156.

13 Glove Workers' Union *v.* Fulton County Gas and Electric Co. (N. Y. P. S. C., 2nd Dist.), P. U. R. 1921 C 789; Customers *v.* Cohoes Power and Light Corp. (N. Y. P. S. C., 2nd Dist.), P. U. R. 1921 D 421.

WATKINS' METER CHARGE WEIGHTED WITH DEMAND COSTS

The weight of a demand component in a meter charge is largely a matter of rate policy. It has been shown above that the meter charge should vary upward from a minimum amount as the investment in meters grows larger. It must be recognized, however, that the meter investment per unit of meter capacity decreases as the size of the meter increases. Consequently, a meter charge varying solely according to the investment in meters reflects total demand costs very inaccurately.

There are grounds, however, for including more than prime customer costs in the meter charge. In effect, the meter charge aims at one of the primary objectives of the demand charge by lessening the quantity charge. But this objective might be more readily reached if the meter charge were graded according to the relative customer demands rather than the relative meter investments. Meter capacity represents maximum demand fairly accurately. Hence, says Watkins,

> This situation might be used as an argument for making the meter charge heavy enough for it to relieve the kilowatt-hour charge to a greater extent than it will when based upon prime cost. It is, at any rate, a strong argument against making the meter charge relatively smaller for the large consumers, at least where the kilowatt-hour rate is at the same time graduated down somewhat with reference to volume of consumption. From this point of view, care should be taken to graduate upwards the meter charge. It should not be allowed to deal merely with the initial cost of the small consumer, on the assumption that it will be relatively negligible for large consumers—until a sufficiently large-sized and otherwise sufficiently distinct type of consumer is reached for them to be entitled to a different type of rate.[14]

But this does not necessarily mean that total demand costs of a given system should be allocated in proportion to meter capacities. Such a procedure might ignore the importance of the diversity and density characteristics of various types of busi-

14 Watkins, *Electrical Rates*, p. 98.

ness. If these were taken into account, the meter charge might be varied according to the meter capacity so as to cover all or a portion of demand costs.[15] The difficulty of administering a graded service charge of this type would not be an insuperable obstacle to its use, for customers could easily be classed according to meter capacity.

It is difficult to ascertain to what extent a service charge based on meter capacity and openly including demand costs has been adopted in this country. Such a charge rarely has been favored by public service commissions, probably owing to the desire to confine the service charge to moderate customer cost elements, and to the conviction that meter capacity may not be an accurate basis for apportioning demand costs among customers of a given class. The Wisconsin Public Service Commission has expressed itself as theoretically in favor of a service charge including demand and customer elements, as compared to one merely covering customer costs. Nevertheless, it has not sanctioned the use of a combined demand and customer charge because of the lack of sufficient data on the time and nature of the customer's demand, and because of the many practical difficulties to be overcome in devising a service charge with a demand component and in applying it to relatively small customers.[16] The Connecticut Public Utilities Commission also has declined to combine the demand component with the customer component, owing to the impracticability

15 Watkins, *op. cit.*, p. 99. A. J. Luick has suggested the use of a "community or service charge" covering customer costs and fixed operating and investment charges, graded according to the size of the meter or capacities required; "Load Building Domestic Gas Rates," *American Gas Journal*, CXXVI (1927), 372. See also W. G. Raymond, *The Public and Its Utilities*, p. 98.

16 *Re* City of New London (Wis. P. S. C.), P. U. R. 1931 E 369; *Re* Wisconsin Public Service Corp. (Wis. P. S. C.), P. U. R. 1933 E 570. It should be noted that in earlier cases, the Wisconsin Railroad Commission had approved "service charges" covering customer and demand cost elements, which however were varied according to connected installation. See, for instance, *In re* Application of La Crosse Gas and Electric Co., 2 Wis. R. C. R. 3 (1907).

of distributing demand charges.[17] Again the Michigan Public Utilities Commission, after carefully determining gas service charges composed of customer costs plus demand costs varied according to meter capacity, reluctantly disapproved them because of consumer protests.[18]

Other Variable Service Charges Recognizing Demand Elements

Watkins' meter charge is novel in that it starts out with prime costs traceable directly and clearly to each customer, which may then be loaded, if it is wished, with as much of the demand costs as is practicable and equitable. But it is by no means unique in recognizing demand elements. To many authorities the service charge means a charge which includes not only customer costs but all or a portion of demand or readiness-to-serve costs as well.[19] In Canada, service charges are often of this type and recognize the varying demands of the consumers.[20] Thus, in 1929, the Rate Committee of the Association of Municipal Electrical Utilities of Ontario approved a proposal to establish increased service charges for domestic service where permanently installed appliances of large capacities were used. The new service charge was of two or more standard

17 Danbury *v.* Danbury and Bethel Gas and Electric Light Co., P. U. R. 1921 D 193.

18 *Re* Grand Rapids, P. U. R. 1923 C 453; *Re* City of Monroe, P. U. R. 1923 D 276. Demand costs were allocated according to the ratio of capacity of each consumer's meter to the total capacity of all the meters attached to the system, upon the assumption that each customer had been provided with the proper size of meter to meet his demands for gas. The cost analyses in these two cases were along the lines suggested by Watkins for electrical meter charges.

19 See p. 64, *supra.*

20 H. D. Rothwell, "Investigation of Domestic Demand," *Ontario Hydro-Electric Power Commission Bulletin*, XVI (1929), 45; "Rate Committee Report of the Association of Municipal Electrical Officials," *Ontario Hydro-Electric Power Commission Bulletin*, XVI (1929), 60, 61; *Report of the Regina Board of Trade Committee on the "Service Charge"* (Regina, 1931), p. 10; Moncton *v.* Moncton Tramways, Electric and Gas Co., Ltd. (New Brunswick B. P. U.), P. U. R. 1932 B 368.

amounts based upon the installed capacity, the meter capacity, and the character of service (2 wire or 3 wire).[21] In England, the service charge also has been made to recognize demand elements, and has been based upon the rateable (tax) value of the property, or the number of rooms wired for electricity, or the wattage of lighting installed.[22]

The essential reason for including demand costs with customer costs in a service charge is the difficulty of attempting to collect the demand costs in a separate charge, distinct from customer and output components. Hence, as large a part of demand costs as can be readily collected in a service charge is so recovered, and the remaining demand costs are covered in the output charge which may be blocked for that purpose.[23] The

21 *Ontario Hydro-Electric Power Commission Bulletin*, XVI (1929), pp. 60-61. The municipalities in the Ontario Hydro System generally use two service charges for domestic service. One is 33c per month per service where the permanently installed appliance load is under 2,000 watts. It should be noted that it is likewise customary to enforce a minimum bill varying from 75c to $2.74. In a few Ontario municipalities the domestic service charge is calculated according to floor area. For commercial and power service, service charges are related to the connected load or measured demand. *Report of the Ontario Hydro-Electric Power Commission*, 1930, pp. 385-396.

22 J. R. Blaikie, "Electricity Supply and the Consumer," *The Electrician* (London), CXX (1928), 409.

23 In the past, the gas industry has paid less attention to problems of the definition, ascertainment, and recovery of demand costs than has the electric industry. In part such neglect is traceable to the fact that experience with the demand characteristics of various types of gas loads has been limited. Furthermore, the problem of demand costs has been considered to be of limited importance for the gas industry, owing to the equalization of demands by storage in holders, and, to a certain extent, in the mains. The difference in demand characteristics of gas and electricity supply is now mainly one of degree. Yet in gas rate-making, the initial or service charge has been preferred to the demand charge. Thus, of the 1249 gas rates in effect in 1933 for general domestic and appliance service (except house heating), only 30 were of a demand type, and those were chiefly Wright demand rates. Of the 249 house-heating rates in effect, 20.1 per cent were demand rates; of the 393 house-heating rates applying to all domestic consumption, 33.1 per cent were demand rates. American Gas Association, Statistical Department, *The Trend of Gas Rate Structures in the U. S. During Recent Years* (N. Y., 1933), pp. 1, 2.

customer cost component may, of course, be merged with the demand cost component in the demand charge. This is the usual procedure in making class rates for large power or lighting in which the demand charge is carefully stated.

A service charge of the demand type is sometimes quoted as a uniform charge per customer, but probably is more often of a type which permits some variation from customer to customer in accordance with the variation in their demand characteristics. In either case, several possible combinations of customer and demand cost elements may have been used to form the service charge. If the service charge is uniform in amount, both the customer and demand components may have been determined to be (a) uniform, or (b) variable. If the demand component has been made equal for all customers, some discrimination may ensue. Thus, if the uniform demand component is fixed by averaging the demand costs of all customers in a given classification, the service charge fails to recover from customers with larger than average demands their share of demand costs, and overcharges those with smaller than average demands. Whether this is a serious discrimination or not depends entirely upon the variation in the demand characteristics among the customers under consideration. This difficulty might be avoided by making the uniform demand component equivalent to the demand cost of customers having the smallest meters, or installations, leaving the remaining demand cost of larger customers to be recovered in other features of the schedule.[24] But this would prove to be equitable only so far as any given customer's quantity of consumption was directly proportional to his responsibility for demand costs. Clearly, the service charge with a uniform per customer demand component escapes serious criticism only if it applies under conditions of uniformity of demand among all customers in a given service classification.[25]

24 *Re* Laclede Gas Light Co. (Mo. P. S. C.), P. U. R. 1929 A 561; Amesbury *v.* Amesbury Electric Light Co., 2 Mass. D. P. U. 66 (1921).

25 Some companies have attempted to apportion some demand costs on

The alternative to a constant demand component in a non-variable service charge is a variable demand component; this means that the customer element would vary inversely with the demand component. Such a charge makes a customer with a small demand pay a higher customer charge than one with a larger demand, and the latter is undercharged either as to customer costs, or as to demand costs. This type of charge scarcely recognizes basic cost variations, unless again there is a practically complete uniformity in customer and demand costs in a given service classification.

It seems necessary, therefore, to vary the combined customer and demand charge in order to recognize variations in demand costs among customers. Such a service charge is most readily applied on the basis of an index of the customer's demand, such as number of rooms or floor space. A basic problem in designing such a rate is to vary the demand component of the service charge as the index of the customer's demand varies, without at the same time varying the customer cost in direct proportion to the demand component. This problem has to be faced even when customer costs are considered to be variable, for they do not necessarily change from customer to customer in accordance with changes in demand costs. Certain other technical problems arise in levying a service charge of the demand type. These involve the method to be used in allocating the demand costs to each type of business, per unit of demand, and the method of measuring or estimating the demand of individual customers. No attempt will be made here to examine or discuss the difficult and extremely controversial problems of demand cost determination, of the recognition of diversity factors, etc. It is arbitrarily assumed that the demand

the basis of the total number of customers served by them, without giving due consideration to differences in class demands. See, for instance, *Re* St. Joseph Gas Co. (Mo. P. S. C.), P. U. R. 1928 B 755, which is an example of the extreme application of the cost analysis illustratively outlined in 1926 by the Rate Structure Committee of the American Gas Association (*cf.* Chap. IV *supra*). See also, Selectmen of Amesbury *v.* Amesbury Electric Light Co., 2 Mass. D. P. U. 66 (1921).

costs for each group of customers are known. The problem of demand costs is considered here only so far as such costs are harnessed with customer costs in the making of service charges.

Forms of Variable Service Charges Including Demand Component

Meter capacity as a standard for a graded service charge has already been discussed in relation to Watkins' meter charge. It has the advantage of being easy to reckon from existing records. But where the majority of customers possess meters of uniform size, it may be discriminatory against those customers whose demands are far below the normal demand. It is difficult to discover to what extent the "meter charge" has been adopted in this country as a service charge. In most instances a uniform service charge, with or without demand elements, is considered to do substantial justice, owing to the prevalence of meters of minimum size.

In addition to meter capacity, all the other accepted methods of applying demand components may be used in formulating a service charge. Measured demand, or its substitutes, such as connected load or active connected load, might be used as a basis of service charge variation. They all, however, have grave defects. Measured demand requires the use of demand indicating instruments or meters which heretofore have been too expensive to install, maintain, and operate for the ordinary domestic customer. Connected load and its modification, active connected load, are troublesome to ascertain, require constant check and calculation to be even fairly accurate, and moreover tend to discourage the installation of additional apparatus by individual customers. For these reasons many companies during the 1920's turned to the so-called room and area methods of applying a demand charge to residential consumers.

Room and Area Service Charges

The essential idea of room-count and floor-area rates is that the actual lighting demand of a residence depends upon the number of rooms or on the floor area, and that the deviations of the actual demand from the demand indicated by the room count or floor area are generally small in individual cases. The relationship between the number of rooms, or floor area, and the energy consumption and demand is said to be fairly constant, except in the case of service for heavy-duty appliances.[26] Hence the demand charge may be varied according to the number of rooms or the floor area so as to yield a revenue sufficient to cover demand and energy costs of lighting in every home served. Under these space rates, the entire lighting demand and consumption of a small householder is billed at the same rate as the entire lighting demand and use of a large dwelling, provided that the use per room or square foot of area is the same. The small householder has the same opportunity to obtain lower follow-on energy rates for large appliance use as does the large householder, without having separate meters for light and power.[27] In contrast, under the block rate, the small householder was discriminated against, and the large one was favored

26 Nash, *P. U. Rate Structures*, p. 50; unsigned article, "Area Charge Essentially a Lighting Demand Charge," *Electrical World*, C (1932), 302-303. Occasionally, the room or area rate is urged for gas service. (B. P. Stockwell, "Promotional Gas Rates," *American Gas Journal*, CXXXVI [1932], 32.) But usually grave doubts are expressed as to whether or not the same relationship holds between home area and gas consumption that exists between electric consumption and home area (L. R. Nash, *op. cit.*, pp. 78-80). However, a recent analysis of a gas business in northern Illinois reveals a positive relationship between gas consumption and home area that is closer than the relationship in the same territory between electric consumption and home area. (See, W. J. Crowley and H. S. Jacobs, "Consumption Characteristics of Gas Consumers," *Journal of Land and Public Utility Economics*, XII (1936), 126-132.) In practice the room or area basis of demand rates is almost unknown for manufactured gas service; the rate lists of the American Gas Association for 1933 show only one instance of a room or area rate for general domestic service, and one instance for house heating service. (American Gas Association, *The Trend of Gas Rate Structures in the U. S. During Recent Years*, 1933, p. 18.)

27 *Re* Alabama Power Co. (Ala. P. S. C.), P. U. R. 1929 A 458, 469; *Re* Alabama Power Co. (Ala. P. S. C.), P. U. R. 1928 E 383. In the latter case the Commission was convinced of "the necessity for graduating such initial charge in order to relieve the customer having a small installation of

because a substantial part of the greater lighting load of the large home fell in the lower rate blocks, even though it had no better load factor than the small home.[28]

The details of the administration of room and area rates differ widely, especially as to the determination of "active" rooms or space. These differences will not be reviewed here. Room or area service charges are sometimes explicit, but frequently they are hidden in a Wright rate. In the latter form of rate, the number of kw.-hr. included in the first one or two blocks is varied in proportion to the active rooms or area; for example, the first 3 kw.-hr. per month per 100 sq. ft. of floor area at 9c. per kw.-hr., next 3 kw-hr. per month per 100 sq. ft. of area at 4c., all excess consumption per month at 2½c. per kw.-hr.[29] In the Wright room or area rate, the demand and customer cost elements are hidden in the unit charge. In the absence of a minimum or customer charge, they are not recovered unless the consumption billed is larger than the blocks in which such costs are buried. Although this defect could be overcome by combining a Wright room rate with a fixed service charge—thus also overcoming the defects of the latter—such a rate usually is regarded as too complicated for general use.

A direct service charge on a room basis may be stated as follows: first 3 rooms or less, 65c. per month, and all additional rooms at 20c. per room per month. An area service

too great a proportion of the fixed service cost which he would necessarily have to pay if an average is struck of all domestic consumers" (p. 385).

28 Unsigned article, "Area Charge Essentially a Lighting Demand Charge," *Electrical World*, C (1932), 302-303. For an account of the experience of the Hartford Electric Light Co. in this respect see, unsigned article, "Area Charge Beneficial to the Customer," *Electrical World*, CIII (1934), 147-148.

29 A room rate on the Wright basis might be as follows:

First 4 kw.-hr. per active room per month at	8c per kw.-hr.
Next 3 kw.-hr. per active room per month at....	4c per kw.-hr.
Excess consumption	3c per kw.-hr.

The number of rooms or the floor area could also be blocked if necessary.

charge could be stated similarly: first 1,000 sq. ft. of area or less, 75c. per month, and all additional area, at 5c. per 100 sq. ft. Direct service charges may be sugar-coated by including with them a small amount of energy, thus stating them as initial charges as in the following example: a service charge of 80c. for the first block of 5 kw.-hr. or less in residences of 3 rooms or less; which is increased by 10c. for each of the next seven rooms. In each of the above cases, the remainder of the rate would consist of the energy component, and possibly an additional demand component to cover the demand costs related to major appliances in excess of the lighting demand.[30]

An explicit service charge on a room basis varies the fixed charge upward from a certain minimum, in accordance with a space index of lighting demand. The stated minimum area or rooms should include the majority of homes, and the variable charge should pertain to large homes. This is usually said to be an equitable arrangement, on the ground that less variation in fixed cost probably occurs among those homes covered by the minimum or fixed service charge than among the larger homes which are subject to additional service charges. Such a

30 The entire demand cost could be included in the service charge, and differences in appliance demand could be recognized by service classifications; this would establish one variable service charge for each classification. Although complex, such a schedule encourages the choice of that classification permitting the most liberal use of appliances at low energy rates. An example of such a schedule is the following, which was established in 1928 by the Hartford Electric Light Company for domestic service:

	Classification		
	Limited Use	Partial Use	Full Use
Flat rate per yr.			
per 100 sq. ft. of floor area	$0.60	$0.96	$1.80
Meter rate per kw.-hr.	0.05	0.04	First 200 – $.03
			Excess – .015

See unsigned article, "Customer Psychology," *Electrical World*, XCIII (1929), 787-789; N. E. L. A. Rate Research Committee, "Electrical Rates and Their Status in 1929," *N. E. L. A. Proceedings*, LXXXVII (1930), 238.

plan successfully meets the test of simplicity in cases where this is important.[31]

Room and area rates sometimes have been preferred to other methods of recovering demand and customer costs on the grounds that they are readily understood, give rise to fewer disagreements as to the correct rating to be applied, permit the customer to install as many outlets as he wishes without increasing the rate, and do away with the annoyance connected with other demand rates in checking the connected load.[32] Another advantage, from the point of view of the public, is that this form of rate may be adjusted for social purposes by favoring the small householder with lower service charges. For that purpose, overhead costs might be allocated on the basis of what the traffic will bear. The value of the premises, determined either by tax assessments or rentals paid, would be even more adaptable for such purposes. However, these objectives, though occasionally recognized in rate structures in England, and in Germany,[33] have never received important recognition in the United States.[34] And it might be urged that, keeping social objectives in mind, the room or area rate fails to consider the fact that income is not always proportioned to the number of rooms or house area inhabited. In fact, if the area or room

31 Nash, *op. cit.*, pp. 164-165; *Re* Alabama Power Co. (Ala. P. S. C.), P. U. R. 1929 A 458. Nash points out that where the success of the rate depends upon revenue produced, the minimum limits of the total service charge should not be made too low, owing to the limited possibilities of stimulating the consumption of small home and apartment customers. The maximum limit for the service charge should recognize the fact that the demands of customers do not continue to increase indefinitely with the size of the home.

32 *Re* Alabama Power Co. (Ala. P. S. C.) P. U. R. 1928 E 383.

33 Batson, *The Price Policies of German Public Utility Undertakings*, pp. 135-136, 171, 175-176.

34 C. S. Reed has said that the primary purpose of a room or area charge is to graduate a fixed charge in proportion to the customer's ability to pay in order to make the fixed charge more saleable, and that the desire to avoid the measurement of demand is only secondary; testimony *In re* Electrical Companies, N. Y. P. S. C., Case No. 6367 (1930-31), *Stenographic Minutes*, p. 2604.

charge is relatively high per unit, it may penalize those who live in spacious houses, for the benefit of those whose premises are smaller, but who are equally able to pay, and whose demand may be equally large. It thus tends somewhat to discourage better housing conditions, and, to the extent that it does so, it may be considered anti-social.

Clearly, if discrimination is to be avoided, the area service charge must be carefully fixed after precise ascertainment of the relationship between rooms or area, and demand and consumption in any given community. The relationship may often turn out to be a rough and ready one, or non-existent, and space rates, therefore, may result in substantial discrimination among customers, unless the charges are quite small.[35] That such a relationship is only approximate, and is only used because it is workable, is evident from the following statement of the Rate Research Committee of the National Electric Light Association in 1929:

> It is not seriously contended that either number of rooms or area, even with adjustments, is directly related to actual demands. Some studies indicated an approximate relationship, but these in turn, have been negatived, in part, by other studies. It is generally conceded, however, that, in spite of some inconsistencies, either number of rooms or area is a reasonable basis of determining the size of initial energy blocks or of independent service charges.[36]

The electric companies serving New York City deemed the room count basis as entirely unsuitable for a new promotional rate, saying, " It would be approximate and inexact for metropolitan conditions even for a residential [lighting] load. . . ."[37]

35 C. O. Fisher, *Some Aspects of Commission Regulation of Public Utilities in Connecticut* (Ann Arbor, Mich.: Edwards Bros., 1933), pp. 44, 98.

36 Report of the Rate Structure Committee, " Electric Rates and Their Status in 1929," *N. E. L. A. Proceedings*, LXXXVII (1930), 237.

37 *Brief for the Four System Electric Companies*, filed by Shearman and Sterling in Case No. 6367, N. Y. P. S. C., February 10, 1931, p. 32. See also, M. S. Sloan, testimony in Case No. 6367, N. Y. P. S. C., *Stenographic Minutes*, p. 1243.

The administration of area and room rates is burdensome; accuracy requires that every customer's billing demand be determined from real estate records or actual inspection of his premises.[38] A vital drawback to room and area rates is that usually they are not well understood or accepted by customers who find them more difficult to check than simpler forms of rates. As a result there is much complaint because of the seeming discrimination between customers of partly similar and partly different characteristics. And where uniformity in rate types is sought, room rates must be ruled out because they are not suitable for commercial customers.[39] Finally, many space rates became obsolete and inaccurate because they were originally founded upon a lighting load and did not take account of changes in demand which came with the widespread introduction of new appliances. The serious difficulties of applying the room and area service charge rates, both from the point of view of the validity of such rates as accurate measures of demand and of their acceptability to the public, have increasingly led to the adoption of simpler forms of rates.[40] This has been especially true where commissions have tried to simplify and standardize rate schedules as in New York and Wisconsin.[41] Public feeling against space rates in New York has expressed itself in a law which prohibits the use of scheduled

38 The electrical companies of New York City considered this an important drawback to the room rate, saying that such a rate "...would cause endless complications if undertaken to be applied to the 1,664,000 residential customers in these boroughs." See, *Brief for the Four System Electric Companies, ibid.*

39 *Ibid.*; Sloan, *loc. cit.*, pp. 226-274, 280, 1243.

40 See, for instance, the following cases: *Re* Public Service Co., 17 N. H. P. S. C. R. 174 (1935); *Re* Green Mountain Power Corp. (Vt. P. S. C.), P. U. R. 1930 B 171; Spiess et al. *v.* Manchester Electric Co. (Conn. P. S. C.), P. U. R. 1933 B 262; Public Service Commission *v.* Southwestern Gas and Electric Co. (La. P. S. C.), 9 P. U. R. (N. S.) 491 (1935); *Re* Arkansas Power and Light Co. (Ark. D. P. U.), 13 P. U. R. (N. S.) 514 (1935).

41 *Annual Report of the Department of Public Service, Public Service Commission*, 1934, p. 23, and 1936, p. 27; *Biennial Report of the Wisconsin Public Service Commission*, 1932-1934, pp. 12-14.

rates and minimum charges after July 1, 1937, whenever these are based in any manner upon the number of rooms, cubic or square foot area, number of outlets, or other similar standards.[42]

The Service Charge Varied According to Customer Density

The view of some rate analysts that customer investment should include a substantial portion of distribution investment, as well as meters and services, raises a question as to the recognition of customer density in arriving at a proper service charge. The limited data available show that the amount of customer cost per consumer, taken on a broad basis, varies considerably in communities which are comparable except for the density of customer population; thus in one town with a dense population, the customer costs were $21.60 per customer per year, whereas, in another town of the same population, which however was widely scattered, comparable customer costs were $33.12 per customer per year.[43] A still wider range was disclosed in calculations by Dr. John Bauer for several sample districts in the territory of the Brooklyn Borough Gas Company. For comparative purposes the seven locations listed below were selected as typical of certain types of residential conditions.[44]

Sample Areas Served by the Brooklyn Borough Gas Company

Group	*Residence Characteristics*
I	Modern apartment house located in apartment section.
II	Apartment house located in section with somewhat different type of apartments; customers shifted more frequently,

42 N. Y. Public Service Law, sec. 66(14), New York Laws 1936, c. 696.

43 N. M. Argabrite, "Fundamental Rate Basis," *Public Service Management*, XLVI (1929), 47.

44 *In re* Brooklyn Borough Gas Co., N. Y. P. S. C., Case 4458 (On Rehearing), *Stenographic Minutes*, pp. 1701-1703.

and used smaller amounts of gas per customer than in District I.

III An industrial section containing multi-family houses.

IV A section of older one family houses of elaborate type.

V A section of one family houses; newer development of rather elaborate type.

VI A section of small one family homes interspersed with two or more family houses; a common type of small house section.

VII A section similar to VI, but with somewhat better houses and with wider frontages.

Data pertaining to these seven locations are summarized in Table 17. Variations in commercial expenses and in maintenance of customer facilities are ignored. These figures do not conclusively prove what the cost variations in Brooklyn were, for they are subject to some criticism as to the representativeness and size of the sample in each group. Nevertheless they do illustrate the wide range in " broad " customer costs on a per customer basis. When, however, such costs are averaged per thousand cu. ft. of gas sold in each location, the range in costs is noticeably narrowed.

On the basis of analyses such as these, Bauer maintains that variations in service conditions in most urban communities make it improper to allocate the cost of mains, services, and any other property, except meters, uniformly to customers. According to Bauer, these property items are determined by the scope of the distribution system, the size of the territory, the variety of demand, and the physical layout; once they are constructed their cost is largely inevitable, regardless either of the amount of use or the number of attached customers.[45] The character of residential consumption has a permanent influence on the costs with regard to the spread of the distribution area. Thus if a single family area is developed, fixed costs per cus-

45 " The Effect of the Service Charge on the ' Small User '," *P. U. Fortnighty*, IX (1932), 272-273.

TABLE 17

BAUER'S ANALYSIS OF COUNTERBALANCING COSTS FOR SELECTED DISTRICTS, BROOKLYN BOROUGH GAS COMPANY, 1929 [a]

	Group						
	I	II	III	IV	V	VI	VII
Physical data							
M. cu. ft. gas sold	1960	1051	889	2258	2039	850	3404
Number of customers [b] .	68	69	40	21	26	36	98
Number of houses	1	1	1	19	29	39	74
Length of mains in ft., (single, 6 inch)	200	165	605	1773	1700	680	1669
Number of services	4	4	10	19	30	40	73
Cost per customer							
Distribution mains [c]	$0.90	$0.73	$ 4.30	$23.97	$19.87	$ 5.74	$ 4.31
Distribution services [c] ..	0.27	0.27	1.15	3.84	4.80	4.71	3.09
Consumers' facilities [d] ..	1.28	1.28	1.28	1.28	1.28	1.28	1.28
Commercial expenses ..	2.87	2.87	2.87	2.87	2.87	2.87	2.87
Consumers' meters	0.86	0.86	0.86	0.86	0.86	0.86	0.86
Total cost per customer	$6.18	$6.01	$10.46	$32.82	$29.68	$15.46	$12.41
Cost per M. cu. ft. of gas sold							
Distribution mains [c]	$0.03	$0.05	$ 0.20	$ 0.22	$ 0.25	$ 0.24	$ 0.12
Distribution services [c] ..	0.01	0.02	0.05	0.04	0.06	0.20	0.09
Customers' facilities [d] ..	0.04	0.08	0.06	0.01	0.02	0.05	0.04
Commercial expenses ...	0.10	0.19	0.13	0.03	0.04	0.12	0.08
Consumers' meters [e]	0.03	0.06	0.04	0.01	0.01	0.04	0.02
Total cost per M. cu. ft.	$0.21	$0.40	$ 0.48	$ 0.31	$ 0.38	$ 0.65	$ 0.35

[a] Source: Exhibits 24-41, *In re* Brooklyn Borough Gas Co., N. Y. P. S.C., Case No. 4458 (On Rehearing).

[b] Active meters at end of year.

[c] Include 10 per cent charge for interest and depreciation, plus maintenance costs.

[d] Include maintenance of installations, work on customers' premises, removing and resetting meters, maintenance of gas appliances, and maintenance of meters.

[e] Represents interest and depreciation.

tomer will remain high, whereas in an apartment area they would be low. These costs are not customer costs, but vary per customer according to conditions under which the service is rendered.[46]

Bauer holds that in a diverse urban territory in which there is a wide diversity of service conditions, there would be a

46 Bauer, *ibid.*, pp. 273-274.

counterbalancing effect if all customer costs were absorbed in the commodity charge. For "as a rule the larger domestic consumers live in sparsely settled districts, in detached or semi-detached single family houses, while the small consumers are usually located in more densely populated districts in large tenement and apartment houses." Thus, if an allocation is made on a per customer basis, the small apartment customer who is served at low cost is saddled with additional costs that are largely due to the large consumers, or conditions under which large consumers are served. "On the other hand, if these costs are apportioned on a commodity basis, there is a counterbalancing effect to lighten the burden of the small consumer, which compensates for the fact that they are less responsible for these costs, per consumer, than the large consumer." At the same time, lower follow-on rates place the large consumer in a favorable position, compared with the small consumer, even where no service charge is found in the rate schedule.[47]

Bauer concedes that there is warrant for the service charge in small communities; there is no particular objection to the collection of a moderate service charge which is co-ordinated with follow-up blocks so as to form a promotional rate schedule. But, it is not essential to a reasonable rate schedule. In fact, where conditions are uniform, the introduction of a service charge may be immaterial from any point of view. Few customers would be affected one way or another, and a really promotional rate, in which the service charge is often said to be essential, may be introduced without a service charge. If there are no great diversities in conditions under which service is rendered, the service charge can be made a practical and equitable part of a rate schedule, but is not essential to such a schedule.[48]

It cannot be denied, however, that the small customers are sometimes served under unfavorable conditions of cost, and

47 Bauer, *loc. cit.*, p. 274; see also Table 17.

48 Bauer, *loc. cit.*, pp. 270, 275, 276; also "When the Service Charge is Discriminatory—and When It is Not." *P. U. Fortnightly*, X (1932), 32-33.

that large customers are often served under favorable conditions. But Bauer points out that no practical rate schedule can possibly meet all differences in cost throughout an urban area.[49] If a service charge is to be made explicit in the rate schedule of a company serving a diverse territory, it should be varied in amount according to the conditions of customer density. However, the great objection to a service charge varying with customer density is that it would lead to excessive complication of rate schedules and probably would not be practicable.[50]

Bauer points out that other rate forms would meet these difficulties. A single straight-line rate would fit the counterbalancing conditions of costs of service in a large territory, for although it would overburden the apartment dwellers with some costs, it would undercharge them as to others, and it would conversely affect the large consumers dwelling in individual homes. A moderate minimum bill could be added to protect the company against losses upon very small users;[51] or perhaps the utility could introduce a block rate with lower prices in successive blocks to promote utilization of service for various purposes.[52] Yet another alternative would be to apply a small service charge, such as 30 cents, which would meet with little or no objection from the public.[53]

The validity of a uniform meter rate based upon the theory of counterbalancing costs depends entirely upon the behavior of costs in any specific area. It does not seem likely that fixed and energy costs for different groups of customers offset each other so well that they can be equitably collected in a uniform meter rate. Even where the conditions under which customers

49 "The Effect of the Service Charge on the 'Small User'," *P. U. Fortnightly*, IX (1932), 276.

50 Bauer, "The Service Charge and Its Development in New York," *National Municipal Review*, XX (1931), 546.

51 *Ibid.*

52 Bauer, *op. cit.*, *P. U. Fortnightly*, IX (1932), 276.

53 Bauer, "New York Law Prohibiting Gas Service Charge Unconstitutional," *National Municipal Review*, XVII (1928), 53.

are served do not vary widely, experience testifies to the usefulness of a rate type which permits recovery of some prime costs in a fixed charge.

The other alternatives to a service charge that are mentioned by Bauer have been widely used. The case of the minimum charge, as against the explicit service charge, has been discussed in the preceding chapter. It is enough to say here that the proposal for the use of the minimum charge indicates some recognition of the existence of fixed costs such as the service charge is designed to meet. Such recognition is explicit in Bauer's proposal for a customer charge of 30 cents or so.

Undoubtedly, if customer costs are viewed broadly, they vary considerably among various districts or groups of customers. But must not causes of cost variation be considered other than those which Dr. Bauer stresses? One fact which may be easily overlooked is that, in the central station industry, densely populated areas are usually served from underground distribution systems that are far more expensive to construct and maintain than the overhead lines that are used in less congested areas. Underground service is usually considered too costly in suburban areas; the suburbanite who wants an underground connection generally must pay for its cost in excess of a normal overhead line.[54] Under these conditions, Nash doubts that the cost of electrical distribution mains per customer would still be smaller in congested areas than in the less populated districts.[55] This objection to Bauer's argument, however, would be valid only if all or most of such underground investment were treated as a customer cost. It is not general practice to do so. Nash himself affirms that the cost of underground systems in excess of a similar length of overhead distribution system is often treated as a general overhead item,

54 L. R. Nash, "The Factor of Discrimination in Service Charges," *P. U. Fortnightly*, X (1932), 272-273. According to one authority, underground distribution systems generally cost from four to five times as much as an all-overhead system; H. B. Gear, "Power Distribution," *Standard Handbook for Electrical Engineers* (New York, 1933), pp. 1545, 1566.

55 Nash, *ibid.*

on the ground that it is a community cost. The remaining investment, equivalent to overhead construction investment, is then apportioned between customer and demand costs.[56] If this is done, Nash's objection loses weight, especially because Bauer does not assign all distribution investment to customer costs.

In part, Bauer's analysis of customer costs is directed to show that the service charges ordinarily collected are far greater than the actual customer costs, if these take customer density into account; hence such service charges are discriminatory. Nash, however, while pointing out the practical difficulty of introducing a customer density differential into a rate schedule, maintains that few existing rate schedules embody full customer costs in their direct or indirect service charges. He supports this statement by saying that his cost analyses have shown customer costs to be not far from $1.00 per month for the average customer; in contrast actual service charges range from 25 cents to $1.00 per month, averaging close to 70 cents. The transfer of certain controversial distribution costs from the customer cost group to the demand cost group might reduce the former, but would not reduce the combined customer and demand costs. Hence, Nash is not convinced that existing service charges cover total fixed costs, including both demand and customer costs, of small residential service, regardless of how great customer density may be.[57]

But this does not effectively answer Bauer's argument. It is highly doubtful whether demand charges can be averaged with customer costs to prove the necessity for a high service charge. Demand costs should be recovered, but not on an average per customer basis without proof of a uniformity in the demand characteristics of, and in the demand investment required for, all small customers. Nash's figures on customer costs are average customer costs, and throw no light upon the

56 *P. U. Rate Structures*, pp. 241-242.

57 Nash, *op. cit.*, *P. U. Fortnightly*, X (1932), 273; *P. U. Rate Structures*, p. 245.

question of variations according to customer density. It may well be that service charges as a whole for domestic consumers are not as high as costs justify. Yet, though Nash admits the merit of recognizing customer density in rate design,[58] he does not show that customers with small customer and demand costs are not discriminated against by a high service charge, even though the latter may be lower than average customer and demand costs.

It is generally conceded that the introduction of a customer density factor into utility rates makes rate structures so complex that they can hardly be administered. Hence density factors, whether of customers or load, have not been distinctly recognized in gas and electric utility rate structures, despite strong arguments advanced in their favor.[59] Exceptions occur in territorial rate differences, or differences in rates for rural as against urban service. The application of density factors within an ordinarily diversified territory would result in a great range of charges which would confuse the consumers and increase the commercial expenses. Such rate structures would have to be changed frequently as the conditions of density changed. At present, the trend of company practice and public policy is toward less rather than more variation in rates; witness the movement toward uniform rate areas in the electric light and power industry. Though a varying charge including distribution cost elements may be justifiable where conditions of customer density and service differ widely and distinctly, it is not usually expedient as a part of a practicable rate schedule.

58 Nash, *P. U. Rate Structures*, p. 245.

59 G. P. Watkins, in 1921, argued the case for recognition of a consumption or load density factor in electrical rates as a means of avoiding existing rate discriminations, and of meeting the competition of isolated plants. He suggested that such a factor should be recognized, in an experimental fashion, through a system of discounts in the energy rate based upon kw.-hr. consumed in a given street block per foot of block front, some account being taken of the number of service connections in the block. But Watkins' suggestion has made little, if any, impression upon rate practice. Watkins, *Electrical Rates*, pp. 162, 172-189, 212, 214.

CONCLUSIONS

No sweeping statement can be made that one type of service charge or another should be used under all circumstances. Whether a service charge should be uniform or variable depends upon the cost elements included in it, and upon the variations in conditions of customer demands and density. A uniform service charge, including customer and demand components, is justifiable under conditions of substantial uniformity of customer demands, metering equipment, and density of population. Such an average uniform service charge would not materially discriminate against most customers. It might be correctly used even where conditions of demand and service varied somewhat, if the charge were limited to the customer and demand costs of customers with the least demand, meter capacity, and length of distribution line. If this were done, however, the avoidance of discrimination would depend upon the design of the remaining portions of the rate schedule.

A varied service charge might frequently be useful in expressing important cost variations, but should not be made too complex for administrative reasons. It should not be designed so as to burden some customers for the relief of others without good reasons, such as the expansion and spread of the service for the benefit of the community as a whole.

CHAPTER VIII

CONCLUSIONS

THIS study was begun in the hope of answering a number of questions on the nature of service charges, as well as the consequences of their use in gas and electric rate schedules. But the questions proved so complex, and the attempt to answer them raised so many new problems that the following conclusions must be largely tentative.

What is the present meaning of the term " service charge "? It is a genus of fixed charges, whose several species are basically intended to recover all or some part of customer costs. Here it contrasts both with a demand charge, and with an energy or commodity charge. This distinction is clear enough. The genus " service charge " includes species and sub-species which also are called " service charges," but which differ markedly from each other, and which, moreover, overlap the genera " demand charge " and " energy charge." Thus the " initial charge " is a species of service charge which includes energy or commodity costs in addition to customer costs. The " minimum charge ", however, does not appear to be a species of service charge, for it is not a fixed charge. Yet it is intended to recover some customer costs, and thus may be said to be a variant of the service charge. The meaning of these terms is clear enough. Unfortunately, however, in practice the service charge may include customer costs alone, or it may also include a part of demand costs. Confusion in meaning would be avoided by referring to the former as a customer charge and to the latter as a service charge. Or, if the term " customer charge " is not palatable to consumers, it could be replaced by service charge, and the term " readiness-to-serve " charge could be used to clearly represent the sense of a combined customer and demand charge. What is needed is a standard definition of terms by an authoritative committee on rate usage and standardization, composed, perhaps, of representatives of the American Gas

Association, the Edison Electric Institute, the National Association of Railroad and Utilities Commissioners, the Federal Power Commission, and similar organizations.

Whatever meaning is finally given to the term "service charge," one question will still remain: what are the customer costs which are basic in any service charge? This does not mean to imply that the service charge should be a rigid embodiment of customer costs, and no more. At times, it may be desirable and expedient to include demand costs in the service charge, in such ways as not to discriminate among customers with different demands. But in most territories it is difficult to collect a fixed charge large enough to cover more than customer costs, and often, where it is attempted, the charge is not properly designed to accord with variations in customers' demands. Without more careful design than has been shown in practice so far, the service charge should be restricted to customer costs.

What are customer costs? There is no need to labor once more the details of the differences between the narrow and the broad views of customer costs. Whether one view is better than the other depends upon factors such as the conditions of service, the characteristics of customers, and the state of development of the market for service in the territories in which the rate is to be applied. On the whole, the narrow view is the more acceptable of the two, and should be used in preference to the broad view. Under ordinary conditions of operation, where there are important differences between the consumption and demands of customers, and in the density of customer population within the rate territory, the narrow view is not likely to be discriminatory, nor is it likely to lead to a service charge so large that customers object to it, even though they do so because they misunderstand it. Yet, there are conditions under which a broad service charge may be justifiable and equitable. These are where there is little variation in density of customer population in the rate territory, and where customers have uniform installations and habits of consumption, so that all are equally responsible for distribution investment and expenses. Under conditions such as these, the fixed charge might be made very large, and the quantity rate low, to stimulate the utilization of excess production and distribution capacity. In fact, the entire rate might then consist of a flat charge, thus making a saving in metering and commercial expenses, but for the fact that there would be wasteful consumption, which would be likely to require costly enlargements of service facilities. Consumers might be willing to pay a high fixed charge comprising most of the bill, if they felt they could use service freely with little or even no additional cost. Such a high charge might be more likely to be accepted than one more moderate but still high, which was accompanied by quantity rates so high as to discourage the liberal use of service. There is ground to believe that the fixed charge should be either very high, so as to permit promotional follow-on rates to be offered, or so low that the customers will not object too much.

As ordinarily applied, the broad view of customer costs is objectionable. It is very likely to be accompanied by arbitrary allocations of costs which rationalize what the company wants to charge as the most profitable rate. The broad view apportions to each consumer costs which cannot be traced to any single customer; in fact, it is consistent with the broad view to allocate a large part of distribution investment and expense to customer costs. The resultant customer cost is bound to be large. If this is divided by the consumption of a typical small customer, it shows that the average rate paid by the small customer is less than the "cost" of serving him. Small customers are served at a loss when they pay less than the out-of-pocket costs of serving them, as they often do; but it cannot be concluded that they are served at a loss when they do not pay the customer "cost" determined on the broad basis. The broad customer cost is likely to be excessive and discriminatory because it averages a large part of distribution cost on a per-customer basis, without allowing for differences in the density of customer population and in the demands of customers. True, most of the costs of distribution are direct costs with respect to the entire group of customers served by the distribution system, but they are not all direct costs of serving the individual customers. To use the number of customers as a divisor of costs that are mainly joint, is to apportion costs arbitrarily.

The reason given for placing as much as possible of the fixed costs of distribution in a service charge is that quantity rates may be thereby reduced. The large customer then can be taken on at a competitive rate which is more than the extra costs of serving him, and the net returns to the company are increased. The low quantity rate also benefits the small customer, though only when he takes more service. There is little objection to low quantity rates that are based on the incremental costs of service, provided that they are correctly ascertained. But there is a strong temptation to underestimate the quantity costs by placing the burden of fixed costs of distribution upon the numerous small customers.

Even "value of service" does not serve to uphold a broad view of customer costs. Few utility companies have been able to collect service charges that cover the full amount of customer costs broadly determined. They have generally found that the "value of service" to a small customer, or what he is willing to pay without any great protest, is not as high as the broad customer cost. Accordingly, utility companies have moderated the service charge or have tried other charges that are more acceptable to the mass of consumers. If, with the moderated service charge in force, small customers still pay less than directly traceable customer costs, they must be subsidized by the larger customers, or else the company must take less than a legal "fair return."

The subsidy to the small user may be justifiable if his ability to pay is less than that of the large consumer. But the few data on the economic status of customers do not show that small consumers are poor as a rule, or that large consumers are usually well-to-do. No rigid conclusion can be drawn as to the relationship between family income and consumption of gas or electricity. The nature of this relationship varies from one service area to another. The data that have been presented (particularly in Appendix D) cast doubt upon the common assumption that small consumers are well-to-do apartment dwellers, and that people with low incomes are among the larger consumers. More information on the purchasing power and standard of living of various kinds of customers is needed before valid conclusions may be drawn as to the desirability of subsidizing small consumers.

It is implicit in the foregoing comments that the theory of gas or electric rates should be stated as a social theory, which aims to maximize the benefits of service to society, and which is not merely a re-statement of a theory of rates aimed at maximizing profits. Utility companies may claim that they give effect to a social theory of rates when they charge according to the "value of service" in order to get business. But "value of service" justifies fixing the service charge at virtually all that the traffic will bear. This amounts to a policy of shifting as much as possible of costs in the form of a service charge to small consumers, to make it possible to reduce rates for large consumption without causing a decline in revenues from small customers. The latter pay the higher charges rather than go without service, and the larger customers are offered more service at low rates largely because they are not willing to pay more for it. This suggests that the making of service charges should not be merely a matter of commercial policy which aims at maximizing net profits; it should also maximize consumers' satisfaction. The difficulty is that the benefits to consumers are measurable in terms of dollars rather than in any concrete unit of gratification. What is needed, obviously, is some sort of social accounting which would enable decisions on matters such as rates to be made on the basis of something better than guesses as to "ability to pay" or "value of service."

What should be done about service charges? Every utility, in formulating its rate schedules, must incorporate an open or hidden charge to recover customer costs. The amount of the charge and the type used depend largely upon the characteristics of the market served. Under ordinary conditions of service, where there is considerable variation in demands and consumption of customers, and in density of population, the customer costs should be limited to those which are directly traceable to the fact of a customer's continuing connection. The minimum of these costs would be interest and taxes on, and depreciation and maintenance of, the meter and the service line or pipe, plus the expense of meter reading, customer accounting, billing, and collecting. In addition, it would be proper to add that estimated share of utilization and general office expenses which arose out of the customer's continued connection, regardless of his demand and consumption. A charge based on these costs would not bear too heavily on small customers, and objections to it would be less than against any other explicit service charge.

In other territories where there is homogeneity of customer demand, consumption, and residential conditions, a larger share of general expenses and a share of distribution expenses and fixed charges could be included without causing discrimination among ratepayers. But a charge of this kind would bear quite heavily on small customers and would be actively opposed by them. If the past relations of the company and the consumers have been none too happy, it probably would be advisable to sugar-coat customer costs, whether broad or narrow, as an initial charge. Where there is a large potential market for domestic service, the service or initial charge could be regarded as a temporary expedient, to be discarded when the consumption of most customers is large enough to assure the company revenues adequate to cover their fixed costs. In such event, a minimum charge could be retained in lieu of an explicit service charge, merely to minimize minor losses on occasional small consumption.

In setting the amount and form of the service charge, the question of promotionality must be kept in mind. The desire to promote revenues must not be allowed to run away with the need for avoiding burdensome rates upon customers who are not able to take advantage of low follow-on rates, either because of habits of living or limited income. Nor should the desire to obtain an immediate increase in sales and revenue be permitted to lead to rates which do not cover full costs of service in the long run. And certainly, the service charge is not always the best means by which rates are made promotional. To use it as a promotional device is to require small customers to shoulder the burden of a change to a new promotional rate and to lighten the losses which the company may be called upon to bear in making the change. The question of promotional rates must be studied not merely with reference to the readjustment of the burden of total costs among various groups of domestic customers, but also with respect to reducing the total cost burden imposed on domestic customers as a whole. This means an examination of ways and means of either shifting a larger share of joint costs to other classes of customers, or of scaling down the company revenues (i. e., operating costs plus fair return) to a lower level, temporarily perhaps, or even permanently if necessary.

APPENDIX A

THE AMOUNT OF CUSTOMER COSTS

THE amount of customer costs per customer may vary from time to time and place to place according to a number of factors. It will be larger if a broad approach to customer costs is followed than if a narrow view is applied. Even if rate-makers agree on the items which are to be included in customer costs, the amount will be affected by the basis of valuing the property, and by the percentage applied to the customer capital items as fixed charges. The factors of density of customer population, efficiency of business management, and fluctuations in prices may lead to substantial variations in the customer cost. In addition, if it is calculated as an average for all customers served by a company, it may differ considerably from the customer cost determined separately for various groups of customers. These factors, however, may at times counteract each other. Hence, it is hardly possible to explain these variations or uniformities accurately.

CUSTOMER COSTS IN THE GAS INDUSTRY

Consider the data in Table 18 on the customer costs of the Brooklyn Borough Gas Company for the year 1926. They emphasize the influence of the use of different valuations, rates of return, and formulae in the determination of customer charges. A change from a present value basis to an investment basis, as shown by the Brooklyn Borough Gas Company's data, substantially decreases customer costs. If the A.G.A. formula is used, as it was by the Company, the method of valuation has a greater influence upon customer costs than a narrow conception, owing to the elimination of the heavy investment in distribution mains under the latter method. Another variable is the rate of return, which is of great importance where a large share of capital investment is allocated to customer cost. If the rate of return is reduced from 8 per cent to 7 per cent, customer costs as determined by the Company are decreased 12 cents per month on the present value basis, and 8 cents on the investment basis.

TABLE 18

CUSTOMER COST PER AVERAGE CUSTOMER OF THE BROOKLYN BOROUGH GAS COMPANY, 1926 [a]

Calculation by:	Basis	Property value	Rate of return applied	Cost per meter month
Company	A.G.A. formula	Present value	8%	$1.84
Company	A.G.A. formula	Adjudicated investment	8	1.48
Wm. Merrifield	Broad view	Investment	8	1.12
John Bauer	Directly traceable costs	Investment	7	.40
Edward W. Bemis	Directly traceable costs	Investment	7	.62

[a] Data compiled from *Brief in Opposition to Change or Modification of Order of July 18, 1929,* submitted to the New York Public Service Commission on behalf of the Brooklyn Borough Gas Company in Case 4458 (on Rehearing), by Whitman, Ransom, Coulson, and Goetz, February 21, 1931, p. 185, and from underlying exhibits.

The major differences in the amount of customer costs are traceable to differences in the conceptions of customer costs. Dr. Bauer takes an extremely narrow view, and does not count as customer costs the following items: return on mains, services, gas appliances, and general office structures; expenses connected with distribution superintendence, new business, and general administration; taxes and depreciation; the major portion of commercial general labor.[1] As a result he obtains a total of 40 cents as a proper customer cost. Mr. Bemis arrives at a somewhat larger figure—62 cents—by making allowances for commercial expenses, work on consumer's premises, distribution superintendence, and return on services. The two figures

1 *Memorandum on Behalf of the Complainants, the Gas Consumers League and the Utility Consumers League of New York*, submitted to the New York Public Service Commission in Case No. 4458 (on Rehearing); Maurice Hotchner, of Counsel; John Bauer, Economic Expert, pp. 12, 14.

probably approximate the total of those costs which are clearly traceable to customers. They may be criticised as being inadequate upon the ground that they have not included allowances for depreciation and taxes on customer investment; on the other hand, Mr. Bemis has included some speculative elements such as general and distribution superintendence.

The customer cost presented by Mr. William Merrifield, an engineer for the Public Service Commission, is the result of an allocation of costs according to his own judgment. It includes practically the same items as does the Company's analysis, but the percentages of items apportioned to customer cost are somewhat smaller than those in the Company's study.

It seems reasonable to conclude that, if customer costs are confined chiefly to those costs directly attributable to each customer, the customer costs per average customer of the Brooklyn Borough Gas Company were about 50 or 60 cents per meter per month. This corresponds closely with the amount of customer costs determined and allowed by the Massachusetts Department of Public Utilities, which has taken a narrow view of customer costs.[2] In contrast, when a broad view of customer costs dominates a rate analysis, as it does in the A.G.A. formula, the customer costs per average gas customer are at least $1.00 per month, and easily run up to $1.50 or more.[3]

The contrasting results of the two methods may be illustrated further. The Brooklyn Union Gas Company claimed a customer cost in 1928 of 96 cents per meter per month; this was derived by applying the A.G.A. formula to property value determined by the adjudicated investment as of 1923 plus additions at cost, and to a rate of return of little more than 5 per cent,

2 Henry C. Attwill, testimony, *Report of the N. Y. Com. Rev. P. S. C. Law* (1930), III, 774-775; Cronin *v.* West Boston Gas Co., P. U. R. 1928 D 181; *Re* Springfield Gas Light Co., P. U. R. 1929 A 229; *Re* Boston Consolidated Gas Company, P. U. R. 1928 C 311.

3 In 1928 the West Boston Gas Company attempted to file a rate on this basis that included a customer's cost charge of $1.90 per month; see Cronin *v.* West Boston Gas Co., P. U. R. 1928 D 181 (Mass.).

which was that actually earned.[4] For the same company in the same year, Bemis, using the same valuation and a return of 7 per cent, found a figure of 57 cents per meter per month by restricting customer costs to those directly traceable to customers.[5] Again, data for the Boston Consolidated Gas Company in 1928 likewise show sharply contrasting results of the two bases of analysis. Customer costs broadly conceived were $1.22 per month, whereas when they were narrowly restricted they amounted to $.62 per customer per month.[6] In a gas rate case which came before the New Jersey Commission in 1929, the A.G.A. method resulted in a customer cost of $1.40, in contrast to $.59 for directly traceable costs.[7]

The average customer costs for domestic customers should be only slightly, if at all, lower than one which is an over-all average for all classes of customers, owing to the predominance of domestic and small customers in most territories. For example, the Brooklyn Borough Gas Company in 1927 had an average of 64,976 meters of all sizes installed in its system, of which 58,906 were small (3 to 5 light) meters. A return of 7% on the average cost of meters and installations for all customers amounted to 87 cents per customer per year, whereas that on the average cost of small, three- and five-light meters was but 9 cents per year less.[8] Differences between average and domestic customers for other items of investment and operating costs are much more difficult to determine. Customer cost analyses should, however, recognize the differences in cost of various groups of customers insofar as possible, especially in instances where there are substantial numbers of commercial and industrial consumers of gas.

4 Exhibits 54, 31, and 36, *In re* Brooklyn Union Gas Co., N. Y. P. S. C. Case No. 4832 (1928).

5 Exhibit 29, *ibid.*; the figure given is an arithmetic mean for a period of four years.

6 Alan Cunningham, testimony of May 1, 1929, Mass. D. P. U. 3530, *Mimeographed Minutes*, pp. 52, 53, 76.

7 *Re* Public Service Electric and Gas Company, P. U. R. 1929 E 21.

8 *In re* Brooklyn Borough Gas Co., N. Y. P. S. C. Case 4458, Exhibit 21, pp. 2-4.

CUSTOMER COSTS FOR ELECTRIC SERVICE

On the basis of many analyses he has made of the cost of electricity for domestic customers, L. R. Nash has stated that customer costs average not far from $1.00 per month for fairly large and efficient properties.[9] He has also stated that the customer cost runs from $1.00 to $1.50 per customer per month, and that this represents "the usual range of experience of normal properties but not the entire range which may be encountered."[10] These figures are based on a rather broad view of customer costs, and include a good share of the fixed capital in overhead distribution, transformers, underground distribution, office facilities, and of the expenses of distribution superintendence, overhead lines maintenance, new business, advertising, plus a large portion of undistributed expenses and fixed charges.[11] These estimates seem to agree with that of F T. Singleton, a member of the Indiana Service Commission, who was of the opinion in 1929 that a reasonable service charge based on customer costs would be approximately $1.00 per month in a city of from 150,000 to 200,000.[12] As in the gas industry, the customer costs of the electric light and power industry are sharply reduced if they are limited to those costs which are directly traceable to the customer.

Recent publications on the costs of distribution of electricity contain data from which the amount and variation of customer costs may be derived.[13] Table 19 shows the customer costs of

9 *Economics of Public Utilities* (1931), p. 458.

10 *P. U. Rate Structures* (1933), p. 244.

11 *Ibid.*, pp. 240-253.

12 *Report of N. Y. Com. Rev. P. S. C. Law,* II, 1518-1519. For gas customers Mr. Singleton places the estimate at $1.25, largely owing to the fact that there are a smaller number of customers over which to spread the total customer cost.

13 The Power Authority of the State of New York, *Report on Cost of Distribution of Electricity* (Albany, N. Y., 1934); M. L. Cooke, ed., *What Electricity Costs* (New York: New Republic, Inc., 1933); Clayton Pike, "Distribution Cost of Electric Energy," Appendix K., *Second Annual Report of the Power Authority of New York* (Albany, N. Y., 1933). This

TABLE 19

CUSTOMER COSTS OF THREE MUNICIPAL ELECTRIC PLANTS, 1932 [a]

	Tacoma, Wash.		Seattle, Wash.		Jamestown, N.Y.	
	Per customer		Per customer		Per customer	
	All	Domestic	All	Domestic	All	Domestic [b]
Number of customers	37,187	32,062	95,195	81,876	15,635	1,493
Customer investment						
Services	$18.40 (Services and Meters)	$12.88 (Services and Meters)	$34.50 (Services and Meters)	$30.55 (Services and Meters)	$ 4.82	$ 5.67
Meters					10.73	10.73
General	.23	.16	.96	.80	1.30	1.65
Total	$18.63	$13.04	$35.46	$31.35	$16.85	$18.05
Fixed charges at 11½% [c]	$ 2.14	$ 1.50	$ 4.08	$ 3.61	$ 1.94	$ 2.08
Operating expenses						
Operation & maintenance ..	$ .63	$ .44	$ 1.25	$ 1.05	$ .31	$ .33
Utilization	.69	.69	.87	.88	.19	.19
Commercial	1.95	1.95	2.43	2.43	1.31	1.31
New business ...	.25	.25	.73	.73	.02	.02
General	.71	.35	1.30	.92	.78	.78
Total	$ 4.23	$ 3.68	$ 6.58	$ 6.01	$ 2.61	$ 2.63
Total customer cost per customer						
Per year	$ 6.37	$ 5.18	$10.66	$ 9.62	$ 4.55	$ 4.71
Per month	.53	.43	.89	.80	.38	.39

[a] Source: New York Power Authority, *Report on Cost of Distribution of Electricity*, pp. 65, 69, 89, 90, 93, 221.

[b] This refers to customers on feeder number 4, which is a residential feeder serving a small single-family dwelling district.

[c] Covers return, depreciation, taxes, and insurance.

three prominent municipal electric plants. The fixed charge percentage of 11½ per cent is one set as a standard percentage by the engineers and economists of the New York Power Authority after making a large number of case studies. Actual fixed charges for these municipal plants would be lower no doubt, as they are in Los Angeles and London, Ontario, as

was written before the publication of the Federal Power Commission's Report called *Cost of Distribution of Electricity*, Power Series No. 3 (Washington: U. S. Government Printing Office, 1936).

shown in Table 20. Variations in density of customer population, and in meter sizes, account for the chief differences in customer investment in Table 19. The high investment and operating expenses per customer in the Seattle municipal plant are largely traceable to the fact that it competes with a private company, and has, therefore, a low customer density and comparatively high commercial and business costs. Of outstanding interest in Table 19 are the low domestic customer costs of the Jamestown and Tacoma systems, which are 39 cents and 43 cents, in contrast to $1.00 which is set by Nash as the lower limit. To some extent, this indicates the results of applying the narrow instead of the broad interpretation of customer costs, although the differences may be traceable in part to other factors such as property valuation, carrying-charge percentages, and operating conditions and policies.

Table 20 shows, for the year 1930, the results of an analysis of customer costs in which actual fixed charges, instead of the standard fixed charges used in Table 19, are applied to customer capital. The figures of 52 cents and 44 cents for the domestic customer costs of the Los Angeles and London (Ontario) systems respectively, are but slightly higher than domestic customer costs of the Tacoma and Jamestown plants. The figure for the Niagara Hudson Company of 84 cents approaches Nash's figure of $1.00, largely because it includes a liberal fixed-charge percentage, as well as a very high allowance for general and utilization expenses.

A summary of customer costs including the same items as Tables 19 and 20, for a number of New York cities in 1932 appears in Table 21. They are not book costs, but instead are determined by applying the standard unit investment and operating costs developed by the engineers of the New York Power Authority to the property and business of each plant. The use of these standard costs probably accounts for most of the consistent uniformity of customer costs from city to city. The application of the unit costs minimizes the effect of excessive

TABLE 20

CUSTOMER COSTS IN THE ELECTRIC LIGHT AND POWER INDUSTRY

	Los Angeles Bureau of Power & Light, 1930 [a]		London,[b] Ontario, 1930	Niagara Hudson Co., 1930 [c]
	Per average domestic customer	Per average customer	Per average domestic customer	Per average domestic customer
Customer investment				
Services	$19.00 (Services and Meters)	$15.10	$ 8.60	$ 8.50
Meters		14.80	13.30	9.87
General	1.12	1.74	2.87	.91
Total	$20.12	$31.64	$24.77	$19.28
Fixed charges—percentage				
Return	4.6%	4.6%	7.0%	[d]
Depreciation	3.0	3.0	2.5	
	7.6%	7.6%	9.5%	13.5%
Fixed charges	$ 1.53	$ 2.40	$ 2.35	$ 2.60
Operating expenses				
Operation, maintenance ...	$ 1.02	$ 1.10	$ .60	$ .87
Commercial	1.48	2.64	1.12	1.87
New business	.92	1.33	.21	.96
Utilization	.66	.93	.24	1.57 [e]
General	.67	.82	.73	2.16
Total	$ 4.75	$ 6.82	$ 2.90	$ 7.43
Total customer cost per customer				
Per year	$ 6.28	$ 9.22	$ 5.25	$10.03
Per month	.52	.77	.44	.84

(a) Data derived from figures cited by T. A. Panter in *What Electricity Costs,* pp. 106, 107, 266-273.

(b) Derived from data cited by E. V. Buchanan in *What Electricity Costs,* p. 113.

(c) Derived from data cited by Morris L. Cooke in *What Electricity Costs,* p. 104-105.

(d) Details not available. Includes taxes.

(e) Includes street lighting expense which was withdrawn in other reports.

valuations of property, and of operating costs which are high because of inefficiency of operation or padding of accounts.

The data in this table show that the effect of giving weight to different meter sizes is quite small. They also illustrate the

TABLE 21

ELECTRIC CUSTOMER COSTS IN NEW YORK STATE, 1932 [a]

Place	Per domestic customer per mo.; 1 meter & service per unit: 2 wire	Per domestic customer per mo.; 1 meter & service per unit: 3 wire	Ratio of services per meter	Per domestic customer per mo.; average meter & service per unit: 2 wire	Per domestic customer per mo.; average meter & service per unit: 3 wire	Per average customer per mo.
Albion	$.67	$.75	.94	$.65	$.72	$.80
Amsterdam	.66	.73	.91	.66	.73	.79
Batavia	.66	.73	.95	.65	.72	.81
Binghamton	.63	—	.75	.59	—	.77
Canajoharie	.63	.69	.77	.62	.68	.72
Corning	.66	.72	.95	.66	.69	.69
Hudson	.67	.74	.96	.66	.74	.76
Malone	.65	.73	.75	.62	.68	.79
Olean	.66	.72	.92	.65	.71	.68
Oneonta	.66	.73	.75	.63	.68	.77
Plattsburg	.66	.73	.95	.65	.72	.80
Poughkeepsie	.62	.69	.50	.56	.61	.72
Utica	.63	—	.86	.61	—	.86
Watertown	.64	.71	.85	.63	.69	.71
Watkins Glen ...	.65	.66	.96	.65	.66	.70
White Plains	.69 [b]		.55	.67 [b]		.72
Consol. Gas Co. .	—	—	.24	—	—	.87
State Average ...	—	—	—	.54 [b]		.84

[a] Sources: New York Power Authority, *Report on Cost of Distribution of Electricity,* Appendices B, C, and G. The average costs for State as a whole are derived from Clayton W. Pike, "Distribution Cost of Electric Energy," Appendix K, *Second Annual Report of the New York Power Authority,* pp. 6-7, Appendices V, X, XV, XIX, XXIII. All figures, except the state-wide average, are based on the standard unit investment and operating costs developed by the Power Authority.

[b] Average domestic customer.

difference between customer costs based upon the assumption that one meter and one service is necessary to serve each customer, and customer costs which recognize the fact that each customer must bear a share of idle meter investment, but, on the other hand, bear no more than his share of a service line which serves several customers. In most cities, consideration of the latter factor does not cause much variation in customer cost, but in several, as is indicated in the percentages of services per active meter in the third column of Table 21, the per customer cost is substantially reduced by a recognition of such customer density. If the customer costs are calculated as an average for all customers, the domestic consumer bears a portion of the larger customer costs of the industrial or commercial user. Generally, as is shown in Table 21, the customer cost per domestic customer is somewhat smaller than that per average customer. In some cities, however, there is little difference between the figures for both bases, owing to the almost exclusively domestic character of the business.

The customer costs in Table 21 may be criticized on the ground that they do not correspond to actuality. In three instances a comparison can be made between an amount representing the per customer costs determined by using actual operating costs per average customer (averaged for the period 1928-1931) plus carrying charges on standard investment, and the per customer cost on the standard cost basis. This comparison is shown below.

	Customer cost per average customer	
Town or Company	Standard unit cost	Actual cost
Malone	$.79	$ .77
Utica	.86	.86
Consolidated Gas Co.	.87	1.17

The similarity of the actual and standard figures for Malone and Utica is very striking. The marked divergence of the two figures for the Consolidated Gas Company is traceable largely to the fact that its commercial costs are much higher than those of other companies.

Data on the direct customer costs of the electric properties in the Consolidated Gas Company system in New York City are summarized in Table 22 below.[14]

14 The companies included are the New York Edison Company, the United Electric Light and Power Company, Bronx Gas and Electric Company, New York and Queens Electric Light and Power Company, and Brooklyn Edison Company, Inc.

TABLE 22

ELECTRIC CUSTOMER COSTS IN NEW YORK CITY, 1933 (a)

Type of Meter	Per meter per month Standard cost basis
5 ampere meter	$.65
6-25 ampere meter	.67
0-25 ampere meter	.66
Average meter (b)	.87

(a) Data from Power Authority of New York, *Report on Cost of Distribution of Electricity,* Appendices C and G.

(b) Includes investment in demand meters.

These figures are based on a city-wide average density of 4.43 meters per service for customers using 25 ampere meters or less, a density of 4.14 meters per service for all customers, and an average of 1.065 meters owned by the company per customer. Because they are based on the standard unit costs of the Power Authority of New York, they are an approximation of what customer costs ought to be rather than what the companies say they are. If actual customer operating expenses, averaged for 1928-1931, are substituted for the standard cost operating expenses, the customer cost per average customer increases from 87 cents to $1.17. These figures, though pertaining to slightly different years, are comparable in the main.

In comparison to the amount of $1.17, the customer cost per average meter as given by the New York City electric companies for 1930 was $1.09.[15] Thus, customer costs per average customer in New York City easily can be shown to to be well over $1.00 per customer, even though a narrow view of customer costs is applied. For domestic customers this figure

15 This does not include the Bronx Gas and Electric Company. The companies' figure excludes new business expenses, but apparently includes all other direct customer expenses exclusive of distribution line items, and assumes a rate of return of 6%. This is based on total meters installed; if the average of meters active during the year were used, the figure would be $1.17 instead of $1.09. *In re* Electrical Companies, Exhibit 54a, N. Y. P. S. C., Case 6367 (1930).

should be somewhat less, about 20 cents, judging by the differences between customer costs per small meters and those of average customers, as shown above in Table 22. And if the standard unit cost basis were followed, customer cost for domestic customers in New York City would be about 66 cents per meter per month.

The range of customer costs for domestic and average customers when based on actual costs is quite wide. This is illustrated in the following tabulation.

TABLE 23

SUMMARY OF ACTUAL CUSTOMER COSTS [a]

Territory	Year	Percentage used for fixed charges	Domestic customer	Average all customers
Jamestown, N.Y.	1932	11.5%	$.39	$.38
Tacoma	1932	11.5	.43	.53
London, Ontario	1930	9.5	.44	—
Los Angeles	1930	7.6	.52	.77
Niagara Hudson Co. .	1930	13.5	—	.84
Seattle	1932	11.5	.80	.89

[a] Sources: Tables 19 and 20.

There is a much narrower range in customer costs when they are based on the standard unit costs developed by the Power Authority study. Table 21 shows a spread from 56 cents to 66 cents for 2 wire meters and services, from 61 cents to 74 cents for 3 wire meters and services, and from 70 cents to 87 cents for all customers. But all these customer cost figures, whether based on actual or standard costs, are considerably below Nash's estimates of $1.00 and $1.00-$1.50, as may well be expected.

APPENDIX B

THE EFFECT OF INCREASED APPLIANCE USE UPON CHARACTERISTICS OF DOMESTIC CONSUMPTION

It is a commonplace in rate discussions that the chief hope for lower costs and hence lower rates for domestic service lies in the more intensive use of existing distribution facilities. This does not merely call for an expansion in the *volume* of consumption. Increased volume may not warrant low rates if it accentuates the existing peak demands. What is necessary is to improve the individual and class diversity factors by building up off-peak demands, thereby bettering the daily and annual load factors.

More information is needed upon these matters, but that at hand testifies to the great load-building value of household appliances, especially those used for cooking, water heating, and refrigeration. Consider, for example, the results of a study made public by Harry A. Snow of the Detroit Edison Company in 1928.[1] The average consumption of a group of refrigerators was 542 kw.-hr. per year, the annual load factor for the group was 69.3 per cent, and the diversity between individual meters was 2.8. Analysis of test records for a group of oil furnace motors showed the following facts: average consumption, 258 kw.-hr. per year; daily load factor, 65 per cent; group average annual load factor, 25.4 per cent; diversity on a December day, 1.5 at the time of the morning peak, 2 in the evening. Analysis of the load of a group of electric ranges revealed no pronounced seasonal variation, and an average annual consumption of 1480 kw.-hr. per year. The average demand of a large group of ranges in Detroit was 336 watts per range, as compared to a possible 4700 watts. Though the annual load factor for an individual range was low (4 per cent

1 "Utility Research Indicates Appliance Load Values," *Electrical World*, XCI (1928), 143-147.

or less), that of a large group with a high diversity mounted to 32.2 per cent.[2] Snow estimates that if the load of these electrical appliances (refrigerators, oil furnace motors, and ranges) were superimposed upon an average lighting and small appliance load in Detroit, annual domestic consumption would rise from the average of 540 kw.-hr. to about 2820 kw.-hr. The addition of such appliances would reduce the fluctuations in the daily and annual residential load curves, and would remove all but small seasonal variations in group consumption.

The few other studies that are available tend to show that as appliances are added to the domestic lighting load, consumption increases, and diversity and load factors for the domestic class are improved.[3] When such domestic loads are combined with other loads, the system load factor is usually bettered considerably.[4] It must be recognized, of course, that the effects of adding appliances in any particular territory will depend upon the state of market development, and upon the number and type of appliances in actual use.

Studies such as the foregoing have led competent authorities to speak in glowing terms of the beneficial effects of the use of appliances. Thus one prominent utility operator has asserted, "The fact is not yet realized that the domestic business, instead of having the poorest load factor of any class, is capable of being developed to the very best, excepting only street lighting and the occasional twenty-four hour industrial."[5]

2 These results differ only slightly from the data for characteristic consumption, diversity, and load factors for ranges reported by R. G. Warner (Engineer, Public Utilities Commission of Conn.), *Load Factor in Relation to Rates*, Paper read at Conference of State Utility Commission Engineers, Washington, D. C., 1934, p. 6.

3 J. D. Ross, "The Effect of Cost of Electricity on Use," in Cooke, *What Electricity Costs* (1933), pp. 213-215; "Report of Promotional Rates Committee, Northwest Electric Light and Power Association," *Electrical West*, LXXVI (1931), 583-587; Samuel Ferguson, "Volume and Diversity in Domestic Sales," *N. E. L. A. Proceedings*, XXCVIII (1931), 87-90.

4 R. G. Warner, *loc. cit.*, p. 7.

5 Samuel Ferguson, *loc. cit.*, p. 90; see also, "Report of Committee on Public Utility Rates," *N. A. R. U. C. Proceedings*, XLIII (1931), 95.

The Effect of Gas Appliances upon Demand and Consumption

It is now generally agreed that the most desirable and most available field for greater use of gas is in refrigeration and water heating. The water heating load is especially desirable when the heater used is one that is automatically controlled to permit the storage of hot water in advance of the demand for it. These, however, are yet in the minority.[6] The amount of gas required to operate an automatic water heater ranges from 2000 to 4500 cu. ft. of gas per month (depending upon family size and income), and the amount generally accepted as typical is 3000 cu. ft. The load factor, especially when the heater demand is restricted to a low limit, is very high. A study of a water heater that was limited to a demand of 5 cu. ft. of gas per hour revealed an individual load factor of 86 per cent, and additional data on diversification indicated a probable group load factor of 100 per cent.[7] A load factor of this kind is exceptional.

6 H. M. Henry, "How to Meet Competition in Domestic Water Heating," *Gas Age-Record,* LXXI (1933), 113.

7 E. E. Linburg, "Limited Demand Water Heaters," Report of the Subcommittee on Water Heating, *A. G. A. Proceedings,* XIV (1932), 553-557.

APPENDIX C

THE RELATIONSHIP BETWEEN RATES AND CONSUMPTION

It is a commonly accepted principle of rate-making that after a promotional rate is introduced consumption increases to such an extent that, even at the low rates made available for liberal use, the customers' bills and the company's revenues also increase, and in time return to old, and perhaps even higher, levels.[1] Numerous studies have been made to ascertain the effect of changes in rates upon consumption. One method of indicating the relationship is to plot the average consumption per customer against the average rate, usually the average revenue per unit of output sold, for a number of different towns, cities, or districts for a given period. A curve may be fitted to the points so determined, which may be called a generalized demand curve; within the limits of available observations, the curve so drawn, usually, but by no means always, approaches a rectangular hyperbola.[2] An essentially similar method of analysis,

1 The primary objective of all carefully formulated, or so-called "scientific" rate systems, such as those of Hopkinson, Wright, and Doherty, was to induce an increase in consumption and revenues. Undoubtedly they were effective in doing so when first introduced. Thus the Doherty rate was said to cause a "very material increase" in the consumption of some customers "due entirely to the form of the rate"; again, after the Doherty rate was instituted in Denver in 1901, the consumption of energy increased 93.5% in one year. (See, H. L. Doherty, "Rates," *N. E. L. A. Proceedings*, XXI (1902), pp. 408, 409. But these earlier rates were promotional primarily as to commercial and industrial consumption. Latterly, the promotion of domestic consumption has attracted the major interest of rate-makers.

2 This method was applied to gas rates by W. D. Marks, in *Finances of Gas and Electric Light and Power Enterprises* (Philadelphia, 1903?), and to electric rates in *Practical Rate-Making and Appraisement* (President's House, South Hadley, Mass., 1914), pp. 203-210. Some later applications, with results essentially similar to those obtained by Marks, are: B. H. Williams, "A Study of the Amount of Electric Current Consumed with Special Reference to the Price Charged," *Giant Power* (Harrisburg, Pa., 1925), Appendix C, pp. 309-321; *Report of the Committee Appointed to*

with similar results, is that of presenting two curves, one depicting an array of the highest to the lowest average prices per unit prevailing in a series of towns or districts, and the other, an array of consumption averages for the same towns.[3] When this is done, the conclusion is drawn that as the average price declines, the average consumption increases.

The method of comparing an array of average prices with an array of consumption averages has several serious defects. It fails to take into account the fact that there may be important differences in the prices of competitive substitutes in the several localities; thus, in a town in which gas and coal are not available at moderate prices, electricity will be used to a far greater extent for cooking than it will in a town in which gas and coal are cheap. Yet, even though the rate schedules are identical in the two towns, the average consumption in the former may well be far larger, and the average rate far lower than in the latter, owing to the billing of large consumption in the lower-priced blocks. The average use is also affected by the intensity of merchandising and advertising activities. Hence, the general demand curve does not accurately describe or measure the response of consumption to a reduction in rates; it does not isolate the effect of rate changes; it does not reveal vital information as to the length of time which must pass after a rate cut until revenues rise to any certain level.

Review the National Problem of the Supply of Electrical Energy (the so-called Weir Committee Report), (London, 1926), p. 5, sec. 13; *Report of the Committee on Electricity Distribution* (Ministry of Transport, London, May, 1936), Appendix II, graphs 3 and 4; E. W. Dickinson and H. W. Grimmit, "The Design of a Distribution System in a Rural Area," *Journal of Institute of Electrical Engineers,* LXX (1932), 202-205; P. P. Stathas, "Fundamental Relationships between Price and Consumption of Electric Service," *N. E. L. A. Bulletin,* XIX (1932), 474-476; Edison Electric Institute, *Load and Revenue Forecasting* (Publication No. A 16, New York, July, 1933), p. 6. See also, Samuel Insull, "Some Comments on the Economics of Electricity Supply," *N. E. L. A. Proceedings,* XXCIII (1926), 193-198.

3 John Bauer and John P. Hogan, "Report of the Marketing Board," *Report of the Saint Lawrence Power Development Commission* (Albany, N. Y., 1931), p. 86, Diagrams 11 and 12.

Some of these defects may be remedied by comparing the results of different rates in market areas in which all other important factors are common, such as the period of time, the policy as to the stimulation of consumption, and the characteristics of consumers. B. J. Sickler uses this method in a careful study of 43 small farming towns in the Mississippi Valley, served from the same electrical system. Average domestic revenues per kw.-hr. in various towns ranged from 13 cents to over 23 cents.[4] When Sickler compared the average revenues in the several rate groups into which the towns were divided with the average consumption in those groups, he found that " a one per cent decrease in rates is accompanied by a 3¼ per cent increase in use of residence lighting customers, and 6¼ per cent by commercial lighting customers." [5] But Sickler also shows that as the size of the town increases so does the consumption per customer; hence the relationship of price and consumption does not depend solely upon the price charged.[6]

Another method of ascertaining the relationship between rate reductions and consumption is to study the annals of rates, revenues, and consumption in a given territory. Sometimes, the trends in rates and revenues of one company following a certain rate policy are compared to the trends of another company pursuing a contrasting rate policy.[7] Analyses of this type

4 B. J. Sickler, "Lower Rates—More Business," *Electrical World,* XCV (1930), 888-889.

5 *Ibid.*

6 Furthermore, in the sample studied by Sickler, low rates were related to an intensive development of the market; this was shown by the general tendency for the number of persons served per residence lighting meter to decrease with lesser rates, though the same relationship did not hold as to commercial customers, who took service regardless of price.

7 The rate annals of the Hartford Electric Light Company and of other companies under the management of Samuel Ferguson have been favorite examples. See, Ferguson, "Promotional Domestic Rates," *N. E. L. A. Bulletin,* XVI (1929), 408-414, and "Volume and Diversity in Domestic Sales," *N. E. L. A. Proceedings,* XXCVIII (1931), 87-90; C. O. Fisher, *Some Aspects of Commission Regulation of Public Utilities in Connecticut,* pp. 42-45; Nash, *P. U. Rate Structures,* p. 173. See also, W. J. Donovan,

usually show that the introduction of a promotional form of rate schedule leads to increases in consumption much larger than those occurring under non-promotional types of rates previously or elsewhere in effect. It is sometimes shown that the proportion of small customers diminishes as lower rates induce them to expand their consumption.[8]

An important advantage of the study of rate and revenue annals is that it permits estimates to be made of the length of time required for average revenue per customer to recover from a rate cut. Thus an examination of the effect of successive rate reductions upon the consumption of 800 established customers in Milwaukee (new customers not being considered), led to the conclusion that a period of two years elapses before the stimulating effect of a rate cut is completely manifested. The ratio of the increase in electricity consumption to a decrease in price was roughly two to one; consumption grew 64 per cent as a result of a 29 per cent decrease in rates.[9] Zuck came to an essentially similar conclusion from a study of the rate annals of 40 companies over a period of five years.[10] The analysis grouped the companies as follows:

Percentage of Decrease in Rates:	Percentage of Increase in: Sales	Revenue
6.66%	24.82%	16.27%
12.03	27.84	12.24
16.93	34.68	11.49
22.10	44.84	9.75
28.07	60.43	17.74

"Report of Counsel," *Report of N. Y. Com. Rev. P. S. C. Law*, pp. 111-112; Martin Insull, "Load Building Opportunities Ahead of Us," *N. E. L. A. Proceedings*, XXCVII (1930), 94-96; L. R. King, "How Rate Structure Influences Gas Sales," *A. G. A. Monthly*, IX (1927), 479-481; C. Heston, "How West Coast Developed its Domestic Business," *Electrical World*, XCV (1930), 1165-1168; and many other similar articles describing the rate experience of particular companies.

8 See, for instance, Ferguson, "Promotional Domestic Rates," *N. E. L. A. Bulletin*, XVI (1929), 413, as to Hartford, Conn.

9 Stathas, *loc. cit.*, p. 477.

10 E. S. Zuck, "Rate Reductions and Sales Increases," *Electrical World*, XCIV (1929), 415-416.

Zuck recognized, however, that the "natural rate of increase" in consumption, taken at 5 per cent per year, would be 25 per cent for the five-year period; hence, the rate reduction in the first group reveals merely the effect of a secular increase in consumption.

The results of other studies, dealing with a variety of samples, may be summarized as follows: (a) experience in typical cases shows that use of a lower energy rate with an initial or service charge results in a recovery of the previous annual revenue per customer within sixteen months;[11] (b) a reduction in electric rates in towns served by the system of the Hydro-Electric Power Commission of Ontario "generally antedates the increased consumption by a period of from two to three years, indicating a loss of revenue throughout this period,"[12] (c) "Data available from certain instances where promotional rates have been introduced for domestic service, involving initial revenue reductions of from 10 to 20 per cent, show that when active sales effort accompanies the change there is a prompt stimulation of sales resulting in a restoration of previous revenue per customer in from ten to fifteen months, with a trend indicating a future increase in revenue above the previous normal."[13]

From these and similar studies, the generalization is often made that a reduction in the price of electricity is soon followed by an increase in consumption sufficient to offset, or more than offset, the loss in revenues. But N. R. Gibson, vice president of the Buffalo, Niagara and Eastern Power Corporation, insists that this cannot be accepted as a uniform law which governs the relationship of price and consumption.[14] It has not always

11 Unsigned Article, "Sixteen Months to Recover with Inducement Rate," *Electrical World*, XCVIII (1931), 1084-1086.

12 Bauer and Hogan, *loc. cit.*, *Report of the St. Lawrence Power Development Commission* (1931), p. 86.

13 N. E. L. A., *Report of the Rate Research Committee on the Status of Electric Rates in 1929* (Publication No. 05, October 1929), p. 4.

14 "Rational Distribution of Electric Energy," paper read before the World Power Conference, Washington, D. C., 1936, pp. 24-26.

held true in the past, and even in instances where it has, there is no assurance that it will continue to hold true in the future, regardless of the extent and intensity of development of the prospective market to which the rates apply. Gibson reviews the records of several rate districts, in which the consequences of rate reductions have been dissimilar. These show that the immediate loss of revenue per residential customer, resulting from rate reductions, varied from a negligible to a large percentage of former revenues. Moreover, the period of time which elapsed until average revenue per customer was restored to former levels ranged from a few months to more than three years. In some districts, the average revenue collected per customer had not returned to old levels after a period of four years, despite some growth in consumption.

One great disadvantage of the foregoing methods of measuring the effect of changes in rates upon consumption and revenues is that they usually make no distinction between a rate reduction achieved merely by a horizontal cut in existing rates, and a reduction which occurs in connection with the introduction of a new form of rate, and which becomes an actual reduction only when customers use substantial quantities of service. The effect of the introduction of a service-charge rate is likely to be confused with the effect of a reduction in the revenues. It may be granted that it is difficult to isolate one from the other, inasmuch as frequently the service-charge rate is made effective at the time of a cut in revenues. A few analyses, however, have attempted to measure the promotionality of a service charge rate. C. S. Reed argues that the basis of analysis must be the cost of additional service to each customer, for when a customer buys an appliance, he is concerned largely with the question of how much it will increase his bill. Reed, therefore, measures the relative promotionality of alternative rate schedules by (a) assuming a certain amount of added consumption for each customer, (b) determining the *average* cost of the new consumption under each of the rates to be compared, and (c) making the added cost under the old rate a base of 100 per cent, to which the added cost under the new or alternative rate may be related.[15] A study of the promotional value of two proposed alternative service-charge rates for the East Ohio Gas Company in 1931 yields the following data:[16]

	Average Price of an Added M. Cu. Ft.	Relation of Added Cost Under Present Rate to Added Cost under New Rate
Present Rate	51.6c	100%
Rate No. 1	63.0c	82%
Rate No. 2	37.1c	139%

Reed says that Rate No. 2 is the most promotional of the three rates; its promotionality is 139 per cent of the promotionality of the present rate. This method demonstrates that as the size of the service charge increases, and as the level of follow-on rate declines, the schedule becomes more and more promotional. It is artificial, however, in that it assumes that consumption will increase. It determines which of the rates in question yields the lowest cost per unit of added service, but it does not correlate the actual change in consumption with the change in added cost to the customer, nor does it reveal anything about the quickness of the return of average revenues to desired levels.

Some writers maintain that the average amount which customers will pay for service, at least in the case of electricity, remains a constant. Thus J. D. Ross asserts that the head of a family has in mind a budgetary allowance for electricity

15 C. S. Reed, Testimony for the Alabama Power Company, Docket # 6483, Alabama Public Service Commission, June 10, 1933; also testimony in Case 6367, N. Y. P. S. C. (1930), *Stenographic Minutes*, pp. 2656-2658.

16 C. S. Reed, *Study of the Service Charge Feature of the Rates of the East Ohio Gas Company* (mimeographed, 1931), p. 19. In this case, the "present rate" consisted of a service charge of 50 cents, plus a quantity rate in five blocks, *increasing* progressively from 50 cents per M. cu. ft. to $1.00 per M. cu. ft. Rate 1 was assumed to be a straight-line meter rate of 63 cents per M. cu. ft., subject to a minimum bill of 75 cents per month. Rate 2 was a service charge of $1.15 per month, plus a commodity rate of 35 cents per M. cu. ft. for the first block of 10,000 M. cu. ft., with all additional service at 60 cents per M. cu. ft.

which he does not wish to exceed. "The residence bill is, therefore, largely a constant, a product of psychology." [17] If rates are lowered, and if the customer's monthly bill is thereby reduced, he will increase his use until his bill once more equals its customary amount.[18] Ross goes further than this, however, and intimates that the subjective budgetary allowance for monthly bills for residence service run about the same throughout the country. This amount, of the magnitude of $2.25 per month, "is surprisingly independent of the rate per kilowatt-hour charged." [19] Ross cites the following data on residence customers in Ontario in 1931 to illustrate the constancy of the size of the bill:

	Average Rate per Kw.-Hr.	Average Consumption per Month	Average Bill Per Month
26 Cities			
(Population over 10,000)	1.45c	148.3 kw.-hr.	$2.14
58 Cities			
(Population 2,000-10,000)	1.97	108.1	2.13
205 Villages			
(Population below 2,000)	2.52	78.4	1.98
All 289 Municipalities	1.59c	133.0 kw.-hr.	$2.12

These conclusions are supported by those curves which show the relationship of average consumption to average rates paid, which, within the known limits of observation, approach a rectangular hyperbola.

17 J. D. Ross, "Effect of Cost of Electricity on Use," in Cooke, *What Electricity Costs*, pp. 202-203; see also, H. L. Doherty, "Equitable, Uniform and Competitive Rates," *N. E. L. A. Proceedings*, XIX (1900), 308; R. S. McBride, "The Substitution of Heating Value for Candle-Power as a Standard for Gas Quality," *Proceedings of International Gas Congress*, 1915, pp. 102-103.

18 Professor Robert T. Livingston, after graphically demonstrating the constancy of average revenues, says: "Decreases of revenue merely increase the sales per meter to a point where the revenue per meter remains about the same." See "Expenses and Sales per Customer," *Electrical World*, XCIV (1929), 382.

19 *What Electricity Costs*, p. 203.

If it is true that the average customer's bill remains a constant, regardless of the rate charged per kw.-hr., there would be a fixed limit to the promotional effect of any rate. It would also follow that it would never pay the company to use promotional rates, because the added output must cost the company something. Yet Ross admits that lower rates and cheaper service may cause the customer to increase budgetary allowance for electric service, especially if the greater use of electricity reduces or eliminates the cost of services previously performed by some other method, or gives the customer more leisure.[20] Only this qualification can account for the increase in the use of new appliances at lower rates, of which Ross is fully aware.[21] Yet, the average monthly bill for all customers may remain constant, or increase slowly, for the increased consumption and larger bills of old customers are continuously being offset by the addition of new customers, many of whom consume little service, and thus counteract the effect of the increase in use on the part of established customers.[22]

20 *Ibid.*

21 *Op. cit.*, p. 205; Ross cites data showing that a reduction of rates of 31.9 per cent by the Ontario Hydro-Electric Power Commission in the period of 1914-1931, resulted in an increase of average consumption 6.3 times that in 1914, and a doubling of the average monthly bill. See also, pp. 205-210.

22 *Cf.*, R. T. Livingston, "Expenses and Sales Per Customer," *Electrical World*, XCIV (1929), 382.

APPENDIX D

RELATIONSHIP OF ECONOMIC STATUS OF CUSTOMERS TO POSSESSION OR AVAILABILITY OF UTILITY APPLIANCES AND FACILITIES

TABLE 24 summarizes the data on the availability to tenants of gas and electricity for lighting, cooking, and mechanical refrigeration, according to rental rates, in the cities covered by the Real Property Inventory of 1934.

For purposes of comparison, it is useful to refer to Table 12 above,[1] which gives the distribution of tenants according to rentals paid. Those tenants who pay less than $20 per month, and who account for at least 51.65% of all rental units, constitute a distinct group in comparison to those paying higher rentals. This is especially true for cooking and refrigeration, though the data fail to distinguish between refrigeration by gas or electricity, and refrigeration by central systems or detached units.

One drawback of the data presented in Table 24 is that they fail to allow for those tenants who do not use gas or electricity because it is not available in the territory in which they live. Though this is not a serious defect, inasmuch as the data refer to well-developed urban areas rather than to the country as a whole, it may be avoided by measuring appliance ownership in terms of families which used gas or electricity for some basic purpose such as cooking or lighting. Another defect of the data in Table 24 is that they refer to tenants, who, as a group, are characterized by a smaller degree of saturation of service facilities than are home-owning customers; this is shown in Table 25 below.

Table 26 avoids the defects of Table 24, but it refers solely to the availability of mechanical refrigeration in a group of nine cities.

[1] P. 107, *supra.*

TABLE 24

PERCENTAGE DISTRIBUTION OF RENTAL DWELLING UNITS BY TYPES OF LIGHTING, COOKING, AND REFRIGERATION FACILITIES, BY MONTHLY RENTAL RATES, FOR SIXTY-FOUR U.S. CITIES, 1934 (a)

Monthly rental rate	Percentages of rental units having facilities for								
	Lighting by			Cooking by			Refrigeration		
	Gas	Electricity	Other	Gas	Electricity	Other (b)	Mechanical (c)	Other or none	Not reported
All groups	0.2%	87.5%	12.3%	67.3%	3.0%	29.7%	14.4%	84.6%	1.0%
Under $10	0.5	48.0	51.5	19.7	0.3	80.0	0.4	99.2	0.4
$10–14.99	0.5	84.9	14.6	52.7	0.7	46.6	1.4	97.7	0.9
15–19.99	0.2	96.1	3.7	74.0	1.3	24.7	3.8	95.0	1.2
20–29.99	0.1	98.9	1.0	85.5	3.8	10.7	13.5	85.4	1.1
30–49.99	0.1	99.5	0.4	89.4	7.0	3.6	41.9	57.4	0.7
50–74.99	0.1	99.4	0.5	88.6	8.7	2.7	66.5	32.8	0.7
75–99.99	0.4	99.1	0.5	84.8	11.4	3.8	73.0	26.1	0.9
100 and over ...	0.3	99.0	0.7	85.6	9.9	4.5	70.6	28.3	1.1
Rent unknown and rent free ..	0.4	64.4	35.2 (d)	44.8	1.9	53.3 (e)	9.3	81.1	9.6

(a) Source: S. L. Kedzierski, *Markets for Gas and Electrical Appliances in Residences* (U.S. Bureau of Foreign and Domestic Commerce, Market Research Series No. 11: 1936), Tables III, VIII, and XIII.

(b) Includes other, none, or not reported.

(c) Mechanical refrigeration refers to iceless refrigeration of all kinds, and includes refrigeration provided from a central machine for all dwelling units of a structure, particularly apartments.

(d) Unreported 13.7%.

(e) Unreported 13.9%.

TABLE 25

AVAILABILITY OF GAS AND ELECTRIC FACILITIES FOR TENANTS AND OWNER-OCCUPIERS, IN SIXTY-FOUR U.S. CITIES, 1934 [a]

Facility	All dwelling units	Owner-occupied dwellings	Rental dwelling units
Lighting	100.0%	100.0%	100.0%
Gas	0.2	0.2	0.2
Electricity	90.6	95.4	87.5
Other or none	9.0	4.3	11.9
Not reported	0.2	0.1	0.4
Cooking	100.0%	100.0%	100.0%
Gas	69.4	72.7	67.3
Electricity	3.9	5.3	3.0
Other or none	26.5	22.0	29.4
Not reported	0.2	——	0.3
Refrigeration	100.0%	100.0%	100.0%
Mechanical	17.0	21.0	14.4
Other or none	82.4	79.0	84.6
Not reported	0.6	——	1.0

[a] Source: Kedzierski, *Markets for Gas and Electrical Appliances in Residences,* Tables I, VI, XI. It should be further noted that 38.9% of the mechanical refrigeration facilities in rental dwellings were furnished as a concession which was included, in whole or in part, in the rental; *ibid.,* Table XVIII.

In all cities, despite striking dissimilarities in relative use of mechanical refrigeration, the proportion of customers using mechanical refrigeration increased as income increased, not counting the no-income group, which lived on borrowings, savings, or charity. Hence, in spite of the prevalence of small incomes, the bulk of refrigeration was enjoyed by families with incomes of $1500 per year and over, as is shown more clearly in Table 27 below.

The percentage saturation of electric range use, and of house-heating by gas increases as income increases. Table 28 summarizes the information for three of the nine cities mentioned in Table 26, in which the percentage of all families using electric ranges to all families using electricity for lighting was over one per cent; the distribution among income groups in cities where the percentage was less than one per cent was

TABLE 26

PERCENTAGE RELATIONSHIP OF FAMILIES USING MECHANICAL REFRIGERATION TO FAMILIES USING GAS AND / OR ELECTRICITY FOR COOKING AND LIGHTING, BY INCOME GROUPS, 1933 (a)

Annual income group	Austin, Texas %	Birmingham, Alabama %	Columbia, S. C. %	Trenton, N. J. %	Portland, Maine %
All groups	19.4	22.4	27.3	9.7	17.0
No income	20.5	8.9	13.0	3.5	7.7
$ 1– 499	3.2	3.5	4.3	1.6	6.8
500– 999	4.4	9.9	6.4	4.7	7.7
1000–1499	12.8	24.7	16.4	8.3	11.2
1500–1999	23.0	37.9	39.0	14.9	16.2
2000–2999	33.3	51.5	54.9	22.8	25.4
3000–4999	51.9	72.8	62.2	41.3	44.7
5000–6999	63.9	74.3	63.6	71.9	60.4
7000 and over	53.3	84.6	100.0	70.0	77.7

Annual income group	Racine, Wis. %	Fargo, N. D. %	Salt Lake City, Utah %	San Diego, Calif. %	All nine cities %
All groups	14.5	28.2	26.7	16.1	19.5
No income	4.2	15.7	11.3	9.9	7.9
$ 1– 499	4.9	4.8	5.5	4.2	4.3
500– 999	7.8	11.4	11.9	8.2	8.6
1000–1499	14.0	19.8	25.9	11.2	16.8
1500–1999	24.2	32.1	40.5	20.9	28.1
2000–2999	59.8	47.0	59.7	31.8	42.6
3000–4999	55.5	69.9	75.0	50.9	60.0
5000–6999	82.4	72.7	79.3	52.3	68.1
7000 and over	63.6	85.7	11.8	60.1	76.4

(a) Source: Ada L. Bush, *Consumer Use of Selected Goods and Services, by Income Groups* (U. S. Bureau of Foreign and Domestic Commerce, Market Research Series, Nos. 5.1–5.9: 1935 and 1936).

haphazard.[2] The distribution of gas house-heating facilities by income groups is summarized in Table 29. The wide variation among cities is accounted for largely by the relative availability of a natural gas supply.

2 In the remaining six of the nine cities, the percentage for all income groups ran as follows: 0.0% in Columbia; less than 0.1% in Austin, and Trenton; and 0.1% in Racine, Fargo and San Diego.

TABLE 27

PERCENTAGE DISTRIBUTION OF FAMILIES HAVING GAS AND / OR ELECTRIC SERVICE AND MECHANICAL REFRIGERATION, BY INCOME GROUPS IN NINE U. S. CITIES, 1933 [a]

Income group	All nine cities
All families	100.0%
No income	1.8
$ 1– 499	4.3
500– 999	10.1
1000–1499	17.9
1500–1999	20.7
2000–2999	24.3
3000–4999	15.1
5000–6999	3.7
7000 and over	2.1

[a] Source: Table 26, *supra;* the cities are those listed in Table 26.

TABLE 28

PERCENTAGE OF FAMILIES USING ELECTRIC RANGES TO FAMILIES USING ELECTRICITY FOR LIGHTING, BY INCOME GROUPS IN SELECTED CITIES, 1933 [a]

Annual income group	Birmingham, Alabama	Portland, Maine	Salt Lake City, Utah
All groups	1.3%	1.5%	22.2%
No income	1.1	0.0	13.4
$ 1– 499	0.2	0.9	6.6
500– 999	0.6	0.6	11.5
1000–1499	1.9	1.0	20.1
1500–1999	2.5	1.5	33.4
2000–2999	2.5	3.3	47.9
3000–4999	2.3	4.2	57.3
5000–6999	3.7	0.0	55.2
7000 and over	0.0	2.8	62.2

[a] Bush, *op. cit.*

Other studies of individual localities give information on a wider variety of appliances than do the studies of the Bureau of Foreign and Domestic Commerce, which were summarized above. Consider Table 30 below, which presents the data on appliance ownership, except of electric ranges, in a group of

TABLE 29

PERCENTAGE RELATIONSHIP OF FAMILIES USING GAS FOR HOUSE HEATING TO FAMILIES USING GAS FOR COOKING, BY INCOME GROUPS, IN SELECTED CITIES, 1933 (a)

Income group	Austin, Texas	Birmingham, Alabama	Salt Lake City, Utah	San Diego, Calif.
All classes	87.2%	1.2%	46.3%	45.1%
No income	82.1	1.2	50.0	37.8
$ 1- 499	83.9	0.5	19.5	27.7
500- 999	84.4	1.0	27.4	42.2
1000–1499	87.3	0.7	38.9	44.7
1500–1999	89.9	1.0	52.5	51.0
2000–2999	93.5	0.9	69.5	53.2
3000–4999	83.5	4.1	115.0	65.2
5000–6999	88.6	10.1	112.0	72.3
7000 and over	80.0	7.7	187.5	78.5

(a) Source: Bush, *op. cit.* Data are omitted for those cities in which gas for house-heating was not clearly distinguished from other fuels. The percentages cited for the following cities represent that portion of families which relied upon fuel *other* than coal, wood, or oil for house-heating purposes: Columbia, S.C., 2.3%; Portland, Me., 8.4%; Fargo, N.D., 0.2%. In the two remaining cities gas was classed with wood, coke, and kerosene as *other* fuels; in Racine, Wis., 0.8% and in Trenton, N.J., 0.7% of the customers relied upon such other fuels.

TABLE 30

PERCENTAGE OF CUSTOMERS OWNING ELECTRICAL APPLIANCES IN CHICAGO, BY MONTHLY HOME RENTALS, 1930 (a)

Appliance	All homes	Monthly rental: Under $50	$50 to $75	$75 to $150	$150 or over
Washer	36%	25%	38%	50%	58%
Vacuum cleaner	87	85	87	94	97
Ironer	5	3	5	12	29
Refrigerator	10	1	9	32	68
Flatiron	96	95	96	97	100
Toaster	37	17	41	60	74
Percolator	16	5	19	28	45
Sewing machine	11	4	11	20	55
Radio	53	42	56	63	71
Heater	10	5	11	20	35
Waffle iron	9	2	9	20	42
Heating pad	7	2	7	14	30
Curling iron	4	5	4	8	13

(a) Data cited by H. Zinder, "Competition and Merchandizing," *N.E.L.A. Bulletin,* XVII, (1930), 551; these originally appeared in *The Edison Round Table,* January 15, 1930.

TABLE 31

PERCENTAGE DISTRIBUTION OF APPLIANCE SALES IN APPLETON, WISCONSIN, ACCORDING TO INCOME GROUPS (a)

Appliance	All families	Annual income per family		
		Under $2000	$2000–$5000	$5000–$10,000
Automatic refrigerator				
Small	100.0%	45.7%	46.4%	7.9%
Medium	100.0	52.4	35.1	12.5
Large	100.0	23.2	36.2	40.6
Electric appliances				
Ironing machines	100.0%	19.9%	38.8%	41.3%
Fans	100.0	41.8	42.3	15.9
Heaters	100.0	40.8	44.5	14.7
Ranges	100.0	43.0	37.6	19.4
Percolators	100.0	50.8	38.0	11.2

(a) Time, Incorporated, *Markets by Incomes*, II, pp. 52, 53, 58. By way of comparison with the above, the distribution of family income was as follows:

Incomes less than $2000 per year	71.7% of all families;
Incomes $2000–$5000 per year	22.8% of all families;
Incomes $5000 and over per year	5.5% of all families.

about 5000 homes in Chicago at the beginning of 1930. It shows clearly that a substantially smaller percentage of the lower rental groups owned appliances than did the upper rental groups, except in the case of vacuum cleaners, flatirons, and radios.

More extensive information is provided by an analysis made by Time, Incorporated, of the purchases of 6,000 families living in Appleton, Wisconsin, for the years 1929, 1930, 1931, and 1932.[3] The purchase of various appliances by three broad income groups is shown in Table 31 above. Families with incomes of $2000 per year or more accounted for the major portion of the sales of appliances, with only two exceptions. It is not surprising, moreover, that on the whole, a much larger proportion of the families in the higher income groups purchased gas and electric appliances than those with lower incomes.[4]

3 Time, Incorporated, *Markets by Incomes* (New York: Time, Inc., 1932).

4 *Ibid.*, I, pp. 22, 23, 29. An exception must be made of electric irons.

BIBLIOGRAPHY

The following list is restricted to the more important references that bear directly upon the subject of service and minimum charges. Many sources that have appeared in the footnotes of this study have been omitted from this list in order to make it reasonably compact. The student of public utility rates will find a mine of references in the bibliography compiled by O. E. Norman, which is listed below under his name. The books by G. P. Watkins and H. E. Eisenmenger are still valuable, though somewhat out of date. L. R. Nash's *Public Utility Rate Structures* (1933) is a convenient and excellent source of information on rate practice.

Achinstein, Asher. *Report of the State Board of Housing on the Standard of Living of 400 Families in a Model Housing Project—The Amalgamated Housing Corporation.* Albany, N. Y., 1931.

Adams, Alton D. "Meter Rents and Minimum Rates," *Electrical World,* LV (1910), 525-27.

Agee, H. H. "Effect of Economic Conditions and Social Trends on Rates," *Gas Age-Record,* LXIII (1929), 643-46, 649.

——. "The Problem of the Small Consumer," *Public Utilities Fortnightly,* V (1930), 670-671.

Allison, James E. *Report of St. Louis Public Service Commission to the Municipal Assembly of St. Louis on Rates for Electric Light and Power,* Appendix B, "Analysis of Rate Calculations for Electric Light and Power"; Appendix C, "Rate Calculations for Electric Light and Power." St. Louis, Mo., 1911.

American Gas Association. *Bibliography. See* O. E. Norman, *infra.*

American Gas Association, Committee on Rate Fundamentals. *Better Forms of Rates for Gas.* New York, 1927.

——. Rate Structure Committee. *Reports* published in the *Proceedings* of the A. G. A., and usually issued as reprints. 1919 to date.

——. *The Trend of Gas Rate Structures in the United States, 1925-1930.* New York, 1930.

——. *The Trend of Gas Rate Structures in the United States During Recent Years.* New York, 1933.

Argabrite, N. M. "Fundamental Rate Basis," *Public Service Management,* XLVI (1929), 47-9, 75-6.

Barker, Harry. *Public Utility Rates.* New York: McGraw-Hill Book Co., 1917.

Batson, H. E. *The Price Policies of German Public Utility Undertakings.* London: Oxford University Press, 1933.

Bauer, John. "The Effect of the Service Charge on the 'Small User'," *Public Utilities Fortnightly,* IX (1932), 269-76.

——. "When the Service Charge is Discriminatory—and When It is Not," *Public Utilities Fortnightly,* X (1932), 32-3.

——. "If Power Cost Nothing to Produce—What Would It Cost to the User?," *ibid.*, pp. 374-82.

——. "New York Law Prohibiting Gas Service Charge Unconstitutional," *National Municipal Review,* XVII (1928), 51-3.

——. "The Principle of the Service Charge," *ibid.*, pp. 302-4.

——. "What is a Proper Service Charge?," *ibid.*, pp. 103-4.

——. "The Service Charge and its Developments in New York," *ibid.*, XX (1931), 543-46.

Beckjord, Walter C. "Economics of Gas Distribution," *A. G. A. Proceedings,* XIII (1931), 743-64.

Caine, Walter E., and Zinder, H. "Regulation of Municipal Electric Utility Rates in Wisconsin: 1931-1934," *Journal of Land and Public Utility Economics,* XI (1935), 12-24.

Clark, John M. *Studies in the Economics of Overhead Costs.* Chicago: University of Chicago Press, 1923.

Clark, Walton. "Meter Rents—A Question of Equity and Policy," *American Gas Light Association Proceedings,* IX (1891), 517-28.

Cooke, Morris L. *On the Cost of Distribution of Electricity to Domestic Consumers.* An Open Letter to the Federal Power Commission and the Several State Public Service Commissions. Concord, N. H.: Rumford Press, 1929 or 1930 (?).

——. "Paying Too Much for Electricity," *The New Republic,* LXXIII (1932), 150-52.

——. (editor). *What Electricity Costs.* New York: New Republic, Inc., 1933.

——. *What Price Electricity for Our Homes.* Philadelphia: 1928.

Crowley, W. J., and Jacobs, H. S. "Consumption Characteristics of Domestic Gas Customers," *Journal of Land and Public Utility Economics,* XII (1936), 126-7.

Dauchy, C. H. "Costs in Rate Making," *Electrical World,* XCIII (1929), 733-5.

——. *Development of Scientific Rates.* See Edison Illuminating Company.

Doane, S. E. "High Efficiency Lamps: Their Effect on the Cost of Light to the Central Station," *N. E. L. A. Proceedings,* XXXVIII (1910), 152-70.

Doherty, H. L. "Equitable, Uniform and Competitive Rates," *N. E. L. A. Proceedings,* XIX (1900), 291-321.

Edison Illuminating Company of Detroit. *The Development of Scientific Rates for Electricity Supply.* Being Reprints of Selected Original Rate Papers, Printed for Private Circulation Only, 1915.

Eisenmenger, H. E. *Central Station Rates in Theory and Practice.* Chicago: Drake & Co., 1921.

Electrical World. (Every volume contains numerous discussions of electric costs and rates.)

——. "Area Charge Beneficial to the Customer," CIII (1934), 147-8.

——. "Area Charge Essentially a Lighting Demand Charge," C (1932), 302-3.

——. "Concerning Meter Rents," LV (1910), 502-3.

——. "Customer Psychology," XCIII (1929), 787-9.

——. "Local Chambers Effective in Justifying Demand Charges," CI (1933), 186.

Empire State Gas and Electric Association. *The Service Charge Rate.* New York, 1918 (?).

Erickson, Halford. *Regulation of Public Utilities.* Madison, Wis.: Democrat Printing Company, 1911.

Farnsworth, Helen E. C. *Rate Structures in the Manufactured Gas Industry.* Doctoral Dissertation Filed at Stanford University, 1930.

Ferguson, Samuel. "Volume and Diversity in Domestic Sales," *N. E. L. A. Proceedings,* XXCVIII (1931), 87-91.

——. "Promotional Domestic Rates," *N.E.L.A. Bulletin,* XVI (1929), 409-14.

Fisher, Clyde O. *Some Aspects of Commission Regulation of Public Utilities in Connecticut.* Ann Arbor, Mich.: Edwards Brothers, Inc., 1933.

Forstall, Alfred E. "Methods of Charging for Gas," *American Gas Light Association Proceedings,* XXII (1905), lxxx-xci.

Frueauff, Frank W. "Methods of Charging for Gas," *American Gas Light Association Proceedings,* XXI (1904), cxiii-cxxxix.

Gas Age-Record. Symposium on "Adoption of the Service Charge," XLVIII (1921), 133-6.

Gear, H. B. Discussion of paper by N. R. Gibson, *Edison Electric Institute Bulletin,* I (1933), 91-2; also in *Electrical World,* CI (1933), 750-1.

Gibson, Norman R. "Variable Elements in the Cost of Distribution of Electrical Energy," *Edison Electric Institute Bulletin,* I (1933), 83-90.

——. *Rational Distribution of Electric Energy.* Third World Power Conference, Washington, D. C., 1936.

Glaeser, Martin G. *Outlines of Public Utility Economics.* New York: The Macmillan Co., 1927.

Grand Rapids Gas Light Company, Brief for: Before the Michigan Public Utilities Commission, "*In re*: The Joint Submission to the Commission of the Regulation of the Grand Rapids Gas Light Company by the City of Grand Rapids and the Said Grand Rapids Gas Light Company under the Provisions of Act No. 419 of 1919," (D 1649, February 5, 1923).

Greene, W. J. "A Method of Calculating the Cost of Furnishing Electric Current and a Way of Selling It," reprinted from *The Electrical World* XXVII (1896), 222-3 in *Development of Scientific Rates for Electricity Supply,* pp. 23-30.

Greene, W. J., Schoonmaker, C. F., and Gorton, C. B. "The Cost of Electric Service," *Electrical World,* LXXX (1922), 878-80.

——. "Allocation of Electric Service Costs," *ibid.*, pp. 928-30.

——. "Handling Cost Elements in Fixing Rates," *ibid.*, pp. 1431-32.

Haase, Ewald. "Cost of Service and Rates Based on the Same," Appendix A, Report of Rate Structure Committee, *A. G. A. Proceedings,* V (1923), 129-52.

Hamilton, F. C. "Practical versus Theoretical Gas Rates," *A. G. A. Monthly,* X (1928), 680-84, 718.

Heitzler, A. H. and Finley, G. R. "An Analysis of Consumption and Revenues of Domestic and Commercial Light Customers," *N.E.L.A. Bulletin,* X (1923), 466-70.

Henry, Hall M. "How to Meet Competition in Domestic Water Heating," *Gas Age-Record,* LXXI (1933), 113-8, 139-42.

Herrmann, R. R. "Effect of Rate Changes on Revenue," *Electrical World,* XXCVIII (1926), 112-4.

Hopkinson, John. "On the Cost of Electricity Supply," *Transactions of the Junior Engineering Society,* III, Part 1 (1892), 1-14; reprinted in *Development of Scientific Rates for Electricity Supply,* pp. 7-20.

Jacobs, Nathan B. "Need the Service Charge be Uniform to All Domestic Users?," *Public Utilities Fortnightly,* X (1932), 198-201.

Jones, Eliot, and Bigham, Truman C. *Principles of Public Utilities.* New York: The Macmillan Co., 1931.

Kennedy, Thomas F. "Factors Affecting the Possibilities and Trend of the Domestic Gas Consumption," *American Gas Journal,* CXXVI (1927), 580-5.

King, L. R. "How Rate Structure Influences Gas Sales," *American Gas Association Monthly,* IX (1927), 479-81.

Knapp, L. H. "House-Heating Data and Trend of Rates in New England," *Gas Age-Record,* LXIX (1932), 129-30, 134, 136-7.

Leffler, William S. "A Study into the Prices of Domestic Electric Service," *N.A.R.U.C. Proceedings,* XLI (1929), 444-65.

Livingston, R. T. "How Cost of Distribution Figures in Electric Bills," *Public Utilities Fortnightly,* XIII (1934), 162-70.

Luick, A. J. "Load Building Domestic Gas Rates," *American Gas Journal,* CXXVI (1927), 361-4, 372-5.

Marshall, A. C., and Snow, H. A. "Distribution Costs—Residence Service," *N.E.L.A. Proceedings,* LXXXVIII (1931), 106-14.

Massachusetts Department of Public Utilities. *Report of the Department of Public Utilities on the Expediency of a Service Charge.* The Commonwealth of Massachusetts, House—No. 2—January 1921.

——. *Special Report of the Department of Public Utilities Relative to the Feasibility and Expediency of Prohibiting So-Called Service Charges in Rate Schedules of Gas and Electric Companies.* The Commonwealth of Massachusetts, Senate No. 50, D.P.U. 4693 (1934).

Monty, Austin. "The Problem of the Unprofitable Customer," *Electrical World,* CVI (1936), 3893-94.

——. "Selling the Unprofitable Customer," *Electrical World*, CVII (1937), 402-3.

Mulligan, W. O. "Domestic Meter Reading and Bill Delivery," *Pacific Coast Gas Association Proceedings*, XXXII (1932), 63-70.

Nash, L. R. *The Economics of Public Utilities*. New York: McGraw-Hill Book Co., 2nd ed., 1931.

——. "The Factor of Discrimination in Service Charges," *Public Utilities Fortnightly*, X (1932), 272-3.

——. *Public Utility Rate Structures*. New York: McGraw-Hill Book Co., 1933.

National Association of Railroad and Utilities Commissioners. Reports of Committee on Public Utility Rates and other special committees, in the *Proceedings* of the annual conventions, 1889 to date. See esp., 1929, 1935.

National Electric Light Association. *Proceedings* of the annual conventions, 1885-1932. See especially the reports of the committees on rate research.

——. *Report of the Rate Research Committee*. Publication No. 15, March, 1931.

——. *Report of the Rate Research Committee*. "Electric Rates and their Status in 1929," Publication No. 05, October, 1929.

——. *Serial Report of the Promotional Rates Committee, 1928-29*. "Promotional Rates," Publication No. 289-293, July, 1929.

Neil, C. E. (compiler). "Bibliography on Electric Rate Making With Special Reference to Analysis of Costs and Classification of Customers: 1883-1928," *N.E.L.A. Proceedings*, XXCV (1928), 1647-57.

New York State. *Report and Hearings of the Commission on Revision of the Public Service Commissions Law*. 4 vols., Albany, N.Y., 1930.

New York State, Power Authority of. *Report on Cost of Distribution of Electricity*. Albany, N.Y., 1934.

New York State, Public Service Commission of. *Annual Reports*. See esp. 1931, 1934, 1935.

——, *Stenographic Minutes* of the hearings and exhibits and briefs filed in Case No. 4458 and Case No. 4458 (on Rehearing). *In re* Rates of the Brooklyn Borough Gas Co., 1928-1930. (Also cited as Customers *v.* Brooklyn Borough Gas Co.)

——. *Stenographic Minutes* of the hearings and exhibits and briefs filed in Case No. 4832 and Case No. 4832 (on Rehearing). *In re* Rates of the Brooklyn Union Gas Co., 1928-1930.

——. *Stenographic Minutes* of the hearings and exhibits and briefs filed in Case No. 6367. *In re* Rates and Rate Structures of Various Corporations Supplying Electricity in the City of New York, 1930-1931. (Also cited as *In re* Electrical Companies.)

Norman, O. E. *Bibliography of Rates for Public Utilities* (1815-1930). New York: American Gas Association, 1930, pp. 177; Supplementary edition, 1934, pp. 48.

Ontario Hydro-Electric Power Commission, *Bulletin.* "Rate Committee Report of the Association of Municipal Electrical Officials," XVI (1929), 60-1.

Parshall, A. "The Service Charge," *American Gas Journal,* CXIV (1921), 223, 234.

Prendergast, William A. *Public Utilities and the People.* New York: D. Appleton-Century Company, Inc., 1933.

Ransom, William L. *The Service Charge as a Part of the Rate for Gas.* Paper read at the meeting of the Empire State Gas and Electric Association at Cooperstown, N.Y., June 24, 1921. Largely reprinted in the *American Gas Journal,* CXV (1921), 25-8, 36-40.

Raymond, William G. *The Public and its Utilities.* New York: John Wiley and Sons, Inc., 1925.

Reed, Charles S. "Are We Peculiar?," *A.G.A. Monthly,* VIII (1926), 267-9, 308.

——. "Installing Better Forms of Rates for Gas," *ibid.,* IX (1927), 561-2.

——. "Salve *vs.* Medicine for Curing Rate Ailments," *ibid.,* IX (1927), 589-90.

——. "Make the Unprofitable Customer Profitable," *ibid.,* X (1928), 275-7.

——. "Who the 'Small User' Really Is," *Public Utilities Fortnightly,* X (1932), 72-80.

Regina Board of Trade. *Report of the Committee on the Service Charge.* Regina, Canada, 1931.

Rothwell, H. D. "Investigation of Domestic Demand," *Bulletin,* Ontario Hydro-Electric Power Commission, XVI (1929), 36-47.

Ruggles, C. O. "The Significance of Minimum and Service Charges in Rate-Making for Public Utilities," *Journal of Political Economy,* XXXII (1924), 56-67.

——. "Discrimination in Public Utility Rates," *ibid.,* 190-206.

Saint Lawrence Power Development Commission. *Report.* Albany, N.Y., 1931. See esp. Report of the Marketing Board.

Seavey, Clyde L. "The Disputed Service Charge," *Public Utilities Fortnightly,* VI (1930), 591-97.

South Carolina Power Rate Investigating Committee. *Report on the Electric Utility Situation in South Carolina.* Columbia, S.C., 1931.

Spitzglass, J. M. "Where We Stand in Rate-Making for Gas or Electricity," *Proceedings of the Pacific Coast Gas Association,* XI (1916), 291-368.

Spurr, Henry C. *Guiding Principles of Public Service Regulation,* III, especially Chap. LXI-LXXIII. Rochester, N. Y.: Public Utilities Reports, Inc., 1926.

Stathas, P. P. "Fundamental Relationships between Price and Consumption of Electric Service," *N.E.L.A. Bulletin,* XIX (1932), 473-9.

Stockwell, B. P. "Promotional Gas Rates," *American Gas Journal,* CXXXVI (1932), 32.

Time, Inc. *Markets By Incomes: A Study of the Relation of Income to Retail Purchases in Appleton, Wisconsin.* 2 vols. New York: Time, Inc., 1932.

United States Light and Heat Corp. *v.* The Niagara Falls Gas and Electric Light Co., et al., 47 Fed. (2d) 567 (C.C.A. 2d., 1931), *Transcript of Record.*

Vincent, W. G., Jr. "Analyzing Gas Costs and Sales," *Proceedings of Pacific Coast Gas Association,* XI (1916), 262-90.

Watkins, G. P. *Electrical Rates.* New York: D. Van Nostrand Co., 1921.

Webber, George B. "Measuring Meter Reading Time Requirements," *A.G.A. Monthly,* XVI (1934), 129-30.

Wehrle, George. "Load Factor and its Bearing Upon Gas Distribution Investment," *American Gas Journal,* CXXXV (1931), 43-6.

Wisconsin Public Service Commission. *Biennial Report.* Madison, Wis., 1934.

Wolf, H. Carl. "Electric Rate Structures," *Electrical World,* XXCIX (1927), 861-4.

Wright, Arthur. "Cost of Electricity Supply," *Minutes of Municipal Electrical Association, 1896;* reprinted in *Development of Scientific Rates for Electricity Supply,* pp. 33-51.

Zinder, H., and Caine, Walter E. "Regulation of Municipal Electric Utility Rates in Wisconsin: 1931-34," *Journal of Land and Public Utility Economics,* XI (1935), 12-24.

INDEX

Bei Fragen zur Produktsicherheit wenden Sie sich bitte an:
If you have any questions regarding product safety,
please contact:

Walter de Gruyter GmbH
Genthiner Straße 13
10785 Berlin
productsafety@degruyterbrill.com